why does a ball bounce?

why does a ball bounce?

101 questions you never thought of asking

ADAM HART-DAVIS

FIREFLY BOOKS

A FIREFLY BOOK

Published by Firefly Books Ltd. 2005

First printing

Publisher Cataloging-in-Publication Data (U.S.)

Hart-Davis, Adam.
Why does a ball bounce : 101 questions you never thought of asking/Adam Hart-Davis.
[224] p. : col. photos. ; cm.
Includes index.
Summary: Questions and answers to 101 scientific queries covering air, the earth, water, fire, ice and rain, mathematics, plants, animals and health.
ISBN 1-55407-113-5
1. Science – Popular works. 2. Science – Miscellanea.
3. Questions and answers. I. Title.
500 dc22 Q163.H378 2005

Library and Archives Canada Cataloguing in Publication

Hart-Davis, Adam
Why does a ball bounce? 101 questions you never thought of asking/Adam Hart-Davis.
Includes index.
ISBN 1-55407-113-5
1. Science – Miscellanea. I. Title.
Q173.H37 2005 500 C2005-902946-3

For Rose, with love and thanks

Published in the United States by
Firefly Books (U.S.) Inc.
P.O. Box 1338, Ellicott Station
Buffalo, New York 14205

Published in Canada by
Firefly Books Ltd.
66 Leek Crescent
Richmond Hill, Ontario L4B 1H1

First published in the United Kingdom by
Ebury Press
Random House, 20 Vauxhall Bridge Road
London SW1V 2SA

Editor: Sam Merrell
Designer: Two Associates and Penny Stock

Printed in Singapore

contents

introduction

This book is about my love for photography and science.

I have enjoyed photography since I was given a Brownie Cresta at the age of about 10, and I remember taking a picture of our cat asleep on the bird table. When I was 18 I spent a year doing Voluntary Service Overseas in India, and took hundreds of photographs in that vast and fascinating country. I started using slides then, on my brother's advice, in order to be able to give talks about my travels. In the late 1970s I began to take my photography more seriously, acquiring a 35 mm camera with interchangeable lenses, and 10 years later I began also to use medium-format equipment.

Meanwhile, I have spent my whole working life trying to explain science ideas in words and pictures. My first job was editing science books; I went on to be a researcher and then producer in the science department at a local television network. I now have no job, but I write articles and books, present radio and television programs, take photographs, and all these activities are aimed at illustrating and explaining ideas in science and technology.

Whether the question is "How does a balloon burst?" or "Why do icicles have bubbles up the middle?", I want to tease out the answer. Photographs may pose the question, or they may help to provide a solution, but for me at least they are part of the beauty of science. In this book you will find some truly scientific pictures, some pictures of mundane things that I find scientifically interesting, and some pictures that cannot be called scientific at all, but are included because they give me excuses to tell stories.

Writing this book was a bit scary, because in many of the topics I am way out of my depth and could make terrible mistakes, which is why I enlisted the help of a gang of friends and experts to check most of the text. In spite of their help I expect I have managed to include some errors.

The logical way to write a book like this would be to compile a list of interesting questions, and then set about taking photographs to answer them. I did it the other way round. I looked for good photographs, wrote what I hope are interesting things about them, and then tried to think of sensible questions.

Anyway, I hope you enjoy reading it and looking at the pictures as much as I enjoyed writing it and taking them.

Adam Hart-Davis

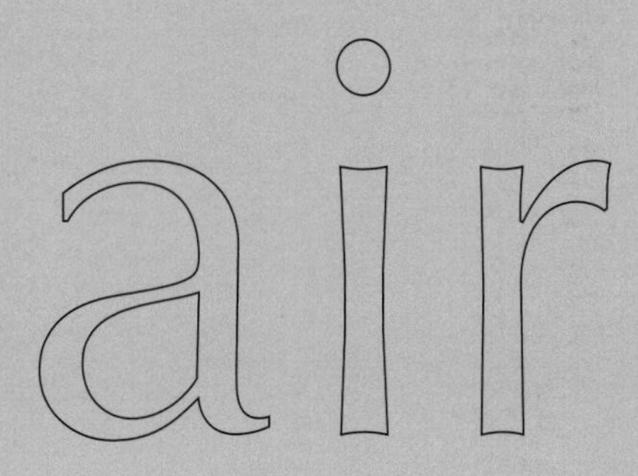

1. How does a balloon burst?

When punctured the balloon goes POP! because the air inside was under pressure; as it escapes, it releases a small shock wave.

Inflating a balloon is hard work because you have to stretch the rubber, which requires energy. When the balloon is punctured, the rubber contracts rapidly to its unstretched state. However, it cannot just collapse inward, because the air in the middle takes time to get out of the way. When you peel an orange, you cannot just squash the skin flat; you have to unwrap it from the outside and then remove the contents. Similarly, the rubber balloon has to unwrap itself from the air inside before it can collapse.

This photograph was taken about 1 millisecond after the moment of puncture. (1 millisecond is $^{1}/_{1,000}$ of a second.) During this time the balloon has split from top to bottom, and the split is rapidly widening. Within about another millisecond the entire skin of the balloon will collect into a twisted wreck at the back.

The duration of this photograph is about 100 microseconds ($^{100}/_{1,000,000}$ or $^{1}/_{1,000}$ of a second). The top of the balloon is sharp, but the moving edges of the rubber are blurred. During this short exposure they have moved about $^{1}/_{3}$ inch (8 mm), which means they are moving at about 525 feet (160 m) per second, or 350 miles (560 km) per hour—half the speed of sound.

Anatomist Karl Langer (1819–87) made pinholes in the skin of corpses, and found they were oval rather than circular, showing similar tension in one direction. He joined the ovals to form what are now called Langer's lines: surgeons generally cut along these lines to minimize scarring after operations.

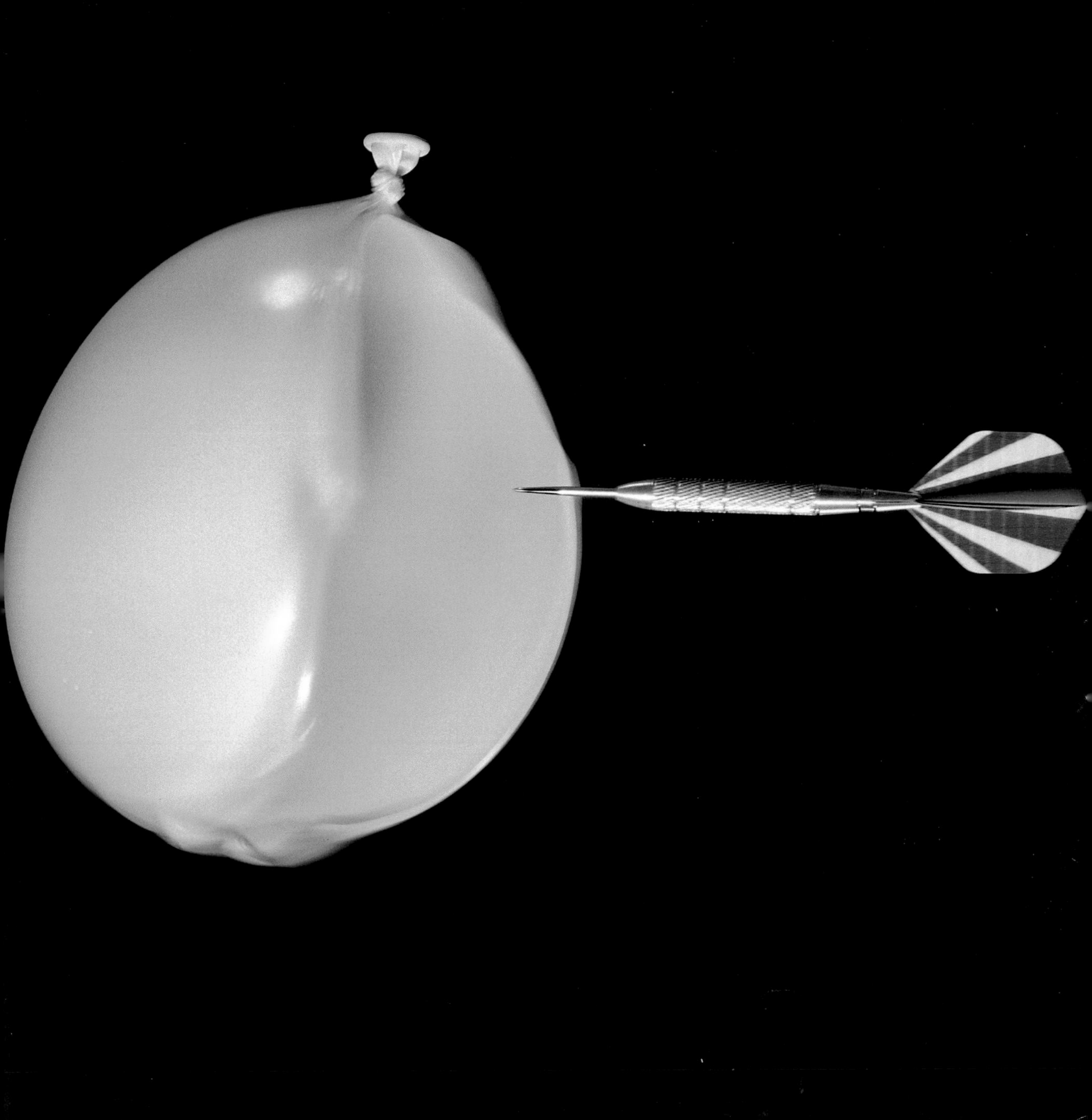

BOLT

2. Why does hot air rise?

The first large hot-air balloons we know about were made in France in 1783 by the Montgolfier brothers Joseph-Michel (1740–1810) and Jacques-Etienne (1745–99). They constructed their balloons from cloth and paper, lit a fire underneath to heat the air and set them free. The first had no cargo; the second carried a sheep, a duck and a chicken, which all survived the flight with no ill effects, and for the third flight they had human volunteers.

What lifts a hot-air balloon is hot gas. When any gas is heated it expands, as long as it is not enclosed in a gas-tight container. When it expands its density is lowered, so any particular volume becomes lighter.

So the air inside an open-necked balloon with a capacity of 1 cubic meter (35 cu. ft.) would weigh about 1 kg (2 lb.) at 20°C (68°F), but only 0.8 kg (1¾ lb.) at 100°C (212°F). Filling it with this hot air would therefore give it a lifting power of 0.2 kg by Archimedes' Principle (see page 63).

In other words hot air rises because when you heat air it expands, which makes a hot-air balloon lighter than the air around it; then cold air rushes in underneath it and pushes it up.

Hot-air balloons have to be big. Human passengers are quite heavy to begin with—some 150 pounds (70 kg) each—and the balloon with its basket, ropes, propane cylinder and burner, weighs about 900 pounds (400 kg). This means that to lift, say, four people and their balloon needs a lift of roughly 1,500 pounds (700 kg).

In practice, a typical balloon is operated with the air inside at about 212°F (100°C), and has a capacity of 124,000 cubic feet (3,500 cu. m). This means the balloon—if it is spherical—has to have a diameter of nearly 65 feet (20 m), which makes it as high as a three-story house.

DID YOU KNOW?

Absolute zero is –273°C (–459°F) or 0 Kelvin. Room temperature is defined as 20°C (68°F) or 293 K (273 + 20). One cubic meter (35 cu. ft.) of air weighs a little more than 1 kilogram (2 lb.) at room temperature. In practice, balloons usually run at about 100°C (212°F) or 373 K, and at this temperature the weight of 1 cubic meter of the air is reduced from 1 kilogram to 0.8 kilogram (293/373).

3. What is the greenhouse effect?

Walk into a greenhouse, and the air inside feels warmer than the air outside. This is partly because, inside, you are sheltered from any wind, but partly because the air inside really is warmer.

DID YOU KNOW?

Light from the sun brings a load of energy to Earth—up to a kilowatt falls on every square meter of surface—which is why in temperate places it feels so pleasant and warm on the skin, and why in hotter places it can burn and cause serious damage.

The sunlight passes straight through the glass of the greenhouse roof and walls, and falls on the ground inside, along with any tables, shelves, plants and so on. The visible light energy is absorbed and converted into heat, as it is on our skin. The warm things inside the greenhouse radiate heat energy, which is infrared light, but this cannot easily pass though glass so it is reflected back by the roof. This means that the energy from the sun is trapped inside the greenhouse as heat, which is why the air inside is warmer than the air outside.

In this greenhouse, on a summer day in a temperate climate, the temperature of the air outside the greenhouse was 19.6°C (67.3°F), while inside it was 32.5°C (90.5°F).

Earth's atmosphere behaves rather like the glass of a greenhouse. Sunlight comes pouring through the air, and when all that visible light energy is absorbed on the surface, it warms up the ground and the oceans. Heat energy is radiated back, but much of it is trapped in the atmosphere by gases that absorb infrared radiation.

Water vapor absorbs infrared, and has always been a greenhouse gas. So has carbon dioxide, although in the last hundred years the amount of carbon dioxide has increased by about 25 percent, probably because we burn so much oil, gas and coal.

Other greenhouse gases include methane, produced in vast quantities by the digestive systems of cows, and chlorofluorocarbons, or CFCs, which for many years were used as coolants in fridges. The more of these gases we allow to escape into the atmosphere, the greater will be the greenhouse effect, and the more probable will be a global warming catastrophe.

OUTDOOR
INDOOR
HIGH ACCURACY DIGITAL THERMO
650-403

AIR
CO2
CH4
-40...120°C
37.6
off
on
Brannan
-40...120°C
39.5
off
on
Brannan
-40...120°C
40.5
off
on
Brannan

4. What are greenhouse gases?

Earth's atmosphere behaves a bit like the glass roof of a greenhouse (see page 14), trapping the energy that comes as sunlight. Greenhouse gases are gases in the atmosphere that help to trap the heat by absorbing the infrared energy that is radiated from the warm earth and oceans.

Two of the best-known greenhouse gases are methane (CH_4) and carbon dioxide (CO_2), and I wanted to find out whether I could demonstrate a greenhouse effect on a small scale. So I took three identical plastic bottles—formerly orange-juice containers—and left one full of air, filled one with methane from the gas supply, and filled the third with carbon dioxide by leaving a piece of dry ice to evaporate in the bottom. Dry ice is solid carbon dioxide, so when a chunk evaporates it generates a lot of carbon dioxide gas, which in this case flushed the air out of the bottle.

I stood the three bottles side by side on a windowsill, so that they would all receive the same amount of sunshine, and measured the temperature of the gas inside. My hope was that both the methane bottle and the carbon dioxide bottle would show higher temperatures than the air bottle, and this indeed was the case. However, at least one high-powered scientist has laughed at me, and said the temperatures are a coincidence, and my experiment is nonsense. So I may just be wrong.

Why should methane and carbon dioxide be greenhouse gases? Why should they absorb infrared radiation? The answer is that they have flexible molecules. In each molecule of carbon dioxide one carbon atom in the center is joined to an oxygen atom on each side; the three atoms are in a straight line. The carbon atom can flap in and out of this straight line, and this bending vibration is triggered by infrared radiation. So if an infrared ray hits a molecule of CO_2, the molecule can absorb the ray and start to vibrate; in effect it is warmed up.

Each methane molecule is made of one carbon atom joined to four hydrogen atoms, arranged around it in a pyramid. This also is a "bendy" molecule that can absorb infrared rays to vibrate in the atmosphere, and so contribute to the warming of planet Earth.

5. How do feathers work?

We are not surprised by feathers, but we should be, since they are one of those wonders of evolution that William Paley (1743–1805) might have used in his argument for a "Designer."

Bishop William Paley, violent antagonist of any suggestion of theories of evolution, wrote in 1802 that if you went walking in the wilderness, you would not be surprised if you found a stone lying on the ground, but you would be surprised if you found a watch, because the presence of the watch implies that there must have been a watchmaker, and there are no watchmakers in the wilderness. So, Paley argued, the complex structures of living creatures could not have arisen by the random process of natural selection; there must have been a Designer—i.e., God.

Feathers keep birds warm and waterproof, and enable them to fly. We poor humans have no equivalent material. Ripstop nylon, used for sails, kites and hot-air balloons, is light and strong, but not very waterproof. Gore-Tex and polyethylene are waterproof but heavy and not warm. Wool and synthetic fleeces are warm, but get horribly wet.

Each feather has a hollow spine of chitinous material (like your fingernails) which is strong because it is a hollow tube. Attached to this spine are lightweight filaments that overlap and interlock like zippers or Velcro. The interlocking filaments trap air between them, and the resulting air-filled knobbly surface is warmly insulating and water-repellent.

When it preens itself, a bird runs each feather through its beak, which both reseals those natural zips and coats the filaments with oil, to make them completely waterproof. If during the trials of daily life some of the zips come undone, or some of the waterproofing is lost, preening will automatically repair the damage.

This flexible system is much tougher and more able to withstand accidents than any human garment. Catch your expensive waterproof coat on a thorn and tear a hole in it, and the coat will leak forever. If a bird brushes its wing on the same thorn, it can repair the damage in seconds—a miracle of evolution.

Paley was wrong. The feather, like so many amazing features of nature, arose only from the relentless processes of natural selection, working over immense periods of time.

6. What keeps the fizz in champagne?

Tear off the foil, unwind the wire and gently ease the cork upward with your thumbs. When it moves you have a choice. Either hold the bottle firmly in one hand and the cork in the other, wait for the pop, keep holding the cork and calmly pour a glass of champagne before it begins to froth out, or hold the bottle in both hands, aim the cork at the moon and let it fly.

Champagne is made like ordinary wine, by fermenting grape juice so that the sugars in the juice are gradually converted by yeast into alcohol and carbon dioxide. Ordinary wine is then bottled and stored, but with champagne the last stage of fermentation is allowed to happen in the bottle, so that carbon dioxide is trapped in with the wine and the gas pressure builds up.

DID YOU KNOW?

The solubility of carbon dioxide gas depends on its pressure; it's more soluble at high pressure. That is why when you release the pressure you immediately make the solution supersaturated, and the gas bubbles out of the solution.

The bottles are thicker and stronger than normal wine bottles, and the cork has to be wired on. When the cork comes out, the gas pressure is released with a pop—like a balloon bursting—and a few seconds later the bubbles of gas released low down in the wine may push some of the fizzy liquid up and out of the bottle.

The same thing happens when you open any bottle of fizzy drink—like cola—especially if it is warm and if you shake it first. However, in those soft drinks, the carbon dioxide is pumped into the liquid under pressure rather than being generated by fermentation.

Once the cork has been released the champagne gradually loses its fizz as the carbon dioxide escapes. Some people claim that a couple of teaspoons poked into the neck of the bottle will prevent the champagne from going flat, but in fact good champagne, kept uncorked in the fridge, will remain fizzy for a couple of days. Even teaspoons made of the best silver are unlikely to have a measurable effect. (A professor friend claims there is only one Law of Champagne: One bottle is never enough.)

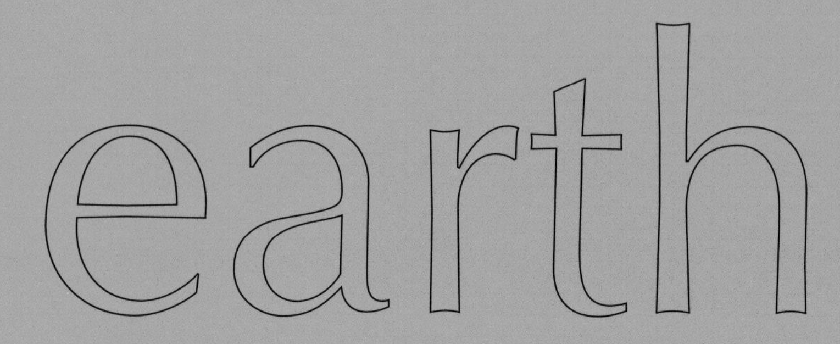

7. How old is the Earth?

In 1650 Archbishop James Ussher of Armagh (1580–1655) used the stories in the Bible to calculate that God had created the world during the fourth week of October, 4004 B.C.

Unfortunately, by the late 18th century, scientific observation did not support this date. Buried in layers of rock were what appeared to be skeletons of animals and fish, which had clearly been dead for a long time. Especially puzzling were those of animals that no longer seemed to exist. Surely if God had created a perfect world in full spate—with lions actually roaring—how could animals have disappeared?

During the 1820s geologist Charles Lyell (1797–1875) traveled around the world studying outstanding geological features, including Niagara Falls. Noting that the falls had receded 50 yards (45 m) in 40 years, he calculated they could have carved out the entire 7-mile (11 km) ravine in 10,000 years.

He also speculated about volcanoes, and in particular Mount Etna, on the island of Sicily. At that time, the catastrophist theory held that such a huge mountain must have been formed in one rapid or catastrophic upheaval of rock. Lyell argued, on the other hand, that continual eruptions of rock over thousands of years could have built the mountain up to its current height.

In 1830 he produced the first volume of a pivotal book: *Principles of Geology*—an attempt to explain the former changes of Earth's surface by reference to causes now in operation. In other words, geology is an active, ongoing process. Earth has been evolving gradually for 4.5 billion years, and is still changing today, as it always has.

8. How are rocks made?

Almost all the rocks on Earth are young compared with the Earth itself, because they are constantly being recycled by erosion and plate tectonics.

There are three main kinds of rock: igneous, metamorphic and sedimentary. Igneous (fiery) is the most common rock (for example, basalt and granite) and is formed by the cooling of hot molten rock from the inside of the Earth. Metamorphic (changed) rock has been metamorphosed by the fierce heat and pressure generated in Earth's crust; for example, shale can be turned into slate and limestone into marble by these processes. For me the most interesting form of rock is sedimentary, which is rock that has settled in layers, rather like the couch, the couch potato and the TV dinner.

Imagine a fierce stream rushing down the side of a mountain and pouring out into a wide lake. As it sloshes down it scours its own bed, wearing away tiny pieces of rock that eventually settle on the bottom of the lake in a layer of mud, sand and gravel. This process may go on for a thousand years, so a thick layer builds up on the lake bed. Then perhaps the stream is diverted, and another one begins to bring silt and rock debris from a different mountain, so that a new layer forms on top of the old, and gradually squashes it. If this happens again and again over thousands of years, the loose sediment gradually gets compressed into solid rock made of layers with slightly varying composition, like a multi-level sandwich. This is sedimentary rock.

Sedimentary rocks often contain fossils, because animal and plant remains get caught up and preserved in the layers of sand and mud. Sometimes a whole layer is made entirely of plant remains—coal—or of the skeletons of tiny sea animals—chalk.

DID YOU KNOW?

Although it often starts at the bottom of the sea, sedimentary rock is sometimes lifted up to great heights by people or by geological movements. Limestone was used to build the pyramids of Egypt, and limestone also forms the peak of Mount Everest, the highest mountain on Earth.

The most common kinds of sedimentary rock are sandstone and limestone. Sandstone is responsible for those spectacular rock formations in the southwestern United States, where all the old Western movies were filmed and where this photograph was taken.

9. How can rocks move?

"Steady as a rock" does not seem to apply at Racetrack Playa in the northwest corner of Death Valley. Drive north 50 miles (80 km) from Furnace Creek, turn left at Ubehebe Crater, carry on for three hours along a dirt road and finally you get to Racetrack. It's a dried-up lake bed, a mile wide by a mile and a half long, surrounded by mountains. There is no tree, no bush, no blade of grass. It's a brown-and-blue world, and even though this is California there is no human habitation for 30 miles (50 km).

Lying about on the hard surface are rocks, varying from pebbles to boulders of perhaps 450 pounds (200 kg); this one is about the size of a couple of bricks—say 10 x 10 inches (25 x 25 cm) on top. The rocks move around on the playa.

Many of them show clear trails where they have been sliding across the surface. Even the biggest seem to move. They move up to 330 feet (100 m) in nearly straight lines, and sometimes two or three seem to have moved in parallel.

Geologists Robert P. Sharp and Dwight L. Carey investigated the phenomenon over a seven-year period. They chose 20 rocks that were already on the playa, painted a number on each, and in the ground 10 feet (3 m) west of each rock planted a stake, with the same number on it. Then every time they visited they could see which rocks had moved, how far and in what direction. Their observations suggested that any one rock moves on average once a year, traveling at about 1.6 feet (0.5 m) per second.

The playa is dead flat, and there are no earthquakes or magnetic anomalies, so how do the rocks move? Could it be a hoax? Well, if someone was moving the rocks, and the surface was soft enough for them to leave trails, then the hoaxer would leave footprints or tire tracks.

One theory was that the rocks are swept along by sheets of ice in cold weather. Sharp and Carey disproved this by enclosing one rock in a corral of stakes—and it moved out.

When it rains, the surface—a particular type of clay—becomes too slippery to walk on. Sharp and Carey reckoned that if under those conditions the wind came howling through the mountains at 90 miles (145 km) per hour, it might just start a rock sliding on a cushion of clay, which would explain why it leaves a trail.

What do you think?

10. What is lava?

Some volcanoes belch smoke, steam and even smoke rings; some make fountains of red-hot rocks and hurl out bombs; some overflow with molten lava, and sometimes the whole top of the mountain blows off.

Lava is the molten rock that many volcanoes spew out. Most of the one hundred Hawaiian Islands in the Pacific Ocean are made entirely of solidified lava. There is a hotspot under the Pacific tectonic plate that forces molten rock up through cracks and holes in the overlying crust to form volcanoes. The lava from these eruptions has built up over thousands of years to form eight major islands, from Kauai in the northwest to the youngest, Hawaii, which is less than a million years old. Called the Big Island, only Hawaii is still close enough to the hotspot to still have active eruptions nowadays, but one of its five volcanoes, Kilauea, is about the most active on Earth.

Kilauea has erupted on and off since it first emerged from the sea more than 50,000 years ago, and continuously since January 1983. Occasionally the eruptions are violent, but more often they are like a pan overflowing, producing some sulfurous gas and masses of lava, which comes in two forms: aa, which is jagged and stony, and pahoehoe, which solidifies into what looks like bundles of ropes. In fact, pahoehoe and aa have the same chemical composition. The lava comes out of the mountain as pahoehoe, but as it cools and becomes more viscous it can break up into lumpy aa.

The active crater of Kilauea moves gradually around the mountain, and the lava flowing down from it has buried around 200 buildings, some roads and even cars under many feet of lava. The red-hot flowing rock sets fire to trees in its path, and produces great clouds of steam when it reaches the sea.

In Italy, the ancient cities of Pompeii and Herculaneum were devastated by the eruption of Vesuvius on August 24, A.D. 79. The mountain first produced a vast column of smoke, and then fountains of ash and pumice stone, which rained down on the surrounding area until it was several yards deep. The following morning there were two tremendous surges of red-hot gas, which flattened the remaining buildings and killed about 3,000 people.

11. What is brimstone?

According to the Bible, "the Lord rained upon Sodom and upon Gomorrah brimstone and fire," and Shakespeare used "Fire and brimstone!" as an expletive, which is curious, since brimstone means sulfur, which is not particularly deadly. However, sulfur burns with a horrible red-yellow flame and choking fumes; the original name was probably "burnstone."

Meanwhile, "brimstone and treacle" was a well-known laxative remedy for constipation, the sulfur working by giving rise to compounds that slightly irritate the digestive tract.

Sulfur is an element, a yellow solid at room temperature, and can appear as powder, as crystals known as flowers of sulfur or as plastic sulfur, which is a rubbery mess.

Alchemists were interested in sulfur for three reasons. First, they were always looking for the possibility of turning things into gold, and sulfur was at least roughly the right color. Second, the fact that it exists in several forms suggested that with the right magical procedure they could turn it into one more form—gold. Third, they sometimes found sulfur lying around on the ground by volcanoes; the photograph here shows sulfur near Kilauea on Hawaii. Volcanoes were thought to be home to gods and magical powers, so anything that came out was of great interest.

In reality, sulfur is ejected from volcanoes because there are both sulfur and sulfur compounds dissolved in the molten magma. Some sulfur is carried by steam, which seeps up through cracks called fumaroles or solfataras. When the steam reaches the surface it drifts away and the sulfur crystalizes on the rocks.

There is sulfur on the moon, on Io, one of the moons of Jupiter and in some meteorites.

DID YOU KNOW?

You eat about a gram of sulfur every day in the form of protein. Sulfur compounds are responsible for the smell and taste of onions and garlic, and the smelliest thing in the world is a sulfur compound called methyl mercaptan.

Sulfur is immensely useful to us today because it vulcanizes rubber, turning sticky latex into useful elastic material; it is converted into sulfuric acid—battery acid—used to make all sorts of products from fertilizer to nylon; and it is a major constituent of gunpowder (see page 73).

12. What is artificial stone?

Originally made for the Lion Brewery, this lion now stands at the east end of Westminster Bridge in London, England, across the river from the House of Commons. I love its leonine snootiness, and the fact that it is in such excellent condition, for the beast has been standing outside in London weather and grime for more than 200 years. It is made of an artificial stone called Coadestone.

Many statues carved from limestone and other natural stone have been severely eroded by wind, weather and especially acid rain, but Coadestone is impervious because it is a ceramic material, like the stoneware used for casseroles. Soft clay, and some secret ingredients, were forced into a mold to make the basic shape. The mold was removed and the detail finished by hand. Then the statue was fired in a kiln to make the finished product.

Eleanor Coade was born in Exeter, England, in 1733, the daughter of a wool merchant who went bankrupt. She moved to London, worked briefly for a draper, and then in 1769 started her own business on the south bank of the river, making the ceramic material that came to be called Coadestone. No better artificial stone has ever been made.

DID YOU KNOW?

The year 1769 was one of great innovation. James Watt (1736–1819) took out his first steam-engine patent; Richard Arkwright (1732–92) took out a patent for a cotton-spinning machine that spun him a fortune; French engineer Nicolas Cugnot (1725–1804) built the world's first steam-powered vehicle; and Wolfgang von Kempelen (1734–1804) constructed a chess-playing machine called The Turk.

Coadestone is so tough that 650 pieces survive, including a river god at Ham House in Surrey, a frieze in Greenwich commemorating Nelson's battles and (so I am told) the entrance to the zoo in Rio de Janeiro. Mrs. Coade must have been a remarkable lady. We know she employed children to push the clay into the molds, because some of the pieces still show tiny fingerprints around the back, but Mrs. Coade was the boss. She had to master the chemistry needed to make the stuff. She must have had or hired artistic skill. And she was clearly a good business-woman, since few women ran companies in the 18th century, and she ran hers successfully for 51 years.

Challenge: Please send me a postcard of the Rio zoo.

13. What is an element?

Today an element is defined as a chemical that cannot be broken down into simpler chemicals, because all its atoms are of the same kind. So hydrogen is an element because all its atoms are hydrogen atoms, and oxygen is an element because all its atoms are oxygen atoms, but water is a combination of hydrogen and oxygen atoms, H_2O, and is therefore not an element, but a compound.

There are more than 100 elements, of which about 90 occur naturally, and the rest have been made artificially. Everything on Earth is made from combinations of these elements.

DID YOU KNOW?

The most common elements are oxygen, silicon, aluminum and iron. Two—bromine and mercury—are liquids and 11 are gases, while the rest are solids. Only a few are found lying around as elements, uncombined: they include carbon, sulfur and gold.

The idea that all matter is built from elementary building blocks goes right back to the ancient Greeks. The man who finally put it together, around 450 B.C., was Empedocles (495–435 B.C.), who lived on the island of Sicily at Agrigento, where this photograph was taken in the valley of the temples. Empedocles said that everything was made of the four elements earth, air, fire and water, combined in various proportions. Some doubters said that air was nothing, but Empedocles proved them wrong. He turned a bucket upside down and shoved it into water. It did not get wet all the way up inside, which proved that there was some physical stuff—air—keeping the water out.

He wanted to prove to his followers that he was immortal; so he took them up to the top of Mount Etna and jumped into the crater of the volcano. He was never seen again, but we remember him to this day, so perhaps his plan was sound.

Later on, Aristotle (384–322 B.C.) took up his scheme of four elements and used it to explain the behavior of matter. Blow bubbles under water, he said, and they rise to get back to their home, the air. Fire goes up toward the sun. And if you drop a stone it falls to Earth because it is made mostly of earth; what's more, it falls faster and faster because as it gets close to home it falls more jubilantly.

14. What is sand?

"Why can't you starve in the desert?" goes the old riddle, "Because of all the sand which is there." Sand covers about a million square miles in the Sahara and over 20 percent of all the deserts in the world. It is composed mainly of particles of silica, or silicon dioxide, in the form of quartz. Quartz is probably the most common mineral on Earth, so it's not surprising there's a lot of it lying around. But "sand" can actually be made of anything, because the word sand describes the size of the particles rather than what they are made of.

Silica is inert; it will not catch fire, soak up water or react with most chemicals. Bricklayers use silica mixed with cement as mortar, and glass manufacturers melt it down as the main ingredient of glass. However, because it does not absorb water, plants will not easily grow in sand, which is one reason why deserts are so barren.

Quartz is one of the hardest rock-forming minerals. Silica sand is therefore tough and rough; sandblasting is a good way of cleaning stone and metal. A stream of sand blown at high speed in a jet of air will rapidly wear away dirt, paint and other unwanted surface material. Sandpaper is stiff paper with a layer of sand glued to one surface. It's used to smooth the surface of wood, because rubbing the wood with sandpaper wears away the surface roughness. This wearing away is called abrasion. You use it to file your nails and to clean your teeth.

DID YOU KNOW?

On some beaches the sand is made entirely of little bits of sea shells. The shells have been broken up by the action of the waves, and the fragments have gradually accumulated on the beach. Sea-shell sand is not silica but calcium carbonate, and would fizz if dropped into acid, because it would release carbon dioxide.

Sand is made by abrasion. Small particles of rock are worn away by weathering—by wind, rain, frost and the erosive action of rivers and the sea. Gradually big rocks get worn down into little pieces. The bits that are soluble get washed away, and what is left are the hard, gritty, quartz fragments that make up sand.

15. Is asbestos dangerous?

Asbestos is a mineral, like chalk or sand, but it comes neither in lumps nor in grains but in fibers—long thin strands that can be twisted into threads and woven into cloth. This can be incorporated into cement to make tiles, boards, pipes and corrugated sheets, which can be used in building.

Because it is a mineral and fully oxidized—in a sense it is already ash—asbestos cannot burn. An asbestos tile in a flame at 1,800°F (1,000°C) will get red hot, but cannot catch fire. Many building materials are susceptible to fire, but not asbestos; walls and roofs made of asbestos are impervious to fire. Therefore, using it in buildings was good for safety. Furthermore the cloth can be fashioned into jackets, gloves and other fireproof clothing—superb for firefighters.

DID YOU KNOW?

Asbestos has been used for thousands of years. Allegedly, Roman café proprietors used it for tablecloths. After customers left, they threw the tablecloth on the fire to burn off the food remains, and then put it back on the table.

Unfortunately, there is a downside, which was known to the Romans, but became horribly apparent in the 1970s. Where the mineral was being processed, the whole factory would fill with dust and tiny fibers of asbestos, floating through the air like snow. The people who worked there began to get seriously ill, wheezing and gasping. The dust and fibers, inhaled from the air, caused intense irritation to the membranes of the lungs, and because asbestos is a mineral it cannot be broken down in the body. Often the disease, asbestosis, developed into a cancer called mesothelioma, and many asbestos workers have died.

The result has been a panic about the dangers of asbestos. It is dangerous if you breathe in masses of fibers or dust floating around in the air. Demolishing buildings and disposing of the asbestos can be hazardous. If you drill into an asbestos sheet with a power drill, or saw through it with a power saw, then you will make clouds of dust, and you have to avoid inhaling it; but asbestos in a tile, in a roof or in cement is unlikely to do you any more harm than an ordinary house brick or a slab of paving stone.

DANGEROUS SUBSTANCE
ASBESTOS WASTE CONTAINS
DANGEROUS SUBSTANCE
DANGEROUS SUBSTANCE

16. When did the penny drop?

"Why should that apple always descend perpendicularly to the ground, thought he to himself. Why should it not go sideways or upwards, but constantly to the earths center? Assuredly, the reason is, that the earth draws it." These are the words of William Stukeley (1687–1765) in his *Life of Newton*.

Isaac Newton (1642–1727) claimed that he first worked out the principles of gravitational attraction at home in Lincolnshire in 1666, and that his ideas were "occasioned by the fall of an apple." In fact, he probably made this up, 60 years later, in order to try to prove that he had solved the problem before his great rival Robert Hooke (1635–1703).

In 1674 Hooke wrote, "All celestial bodies whatever have an attraction or gravitating power towards their own centres ... " and so on. Indeed, he wrote then most of what have come to be called Newton's laws of motion. Newton wrote nothing about gravity or laws of motion until more than 10 years later.

Newton, however, knew he was always right. He was next to God. No one else could possibly beat him to a scientific idea. So he simply stole Hooke's results and wrote them as his own. He did go further than Hooke, however, because he managed to demonstrate the inverse square law: that the force of gravity between two masses decreases in proportion to the square of the distance between them. Hooke failed to do this.

Some say that Robert Hooke was the greatest experimental scientist of all time. He was certainly immensely ingenious and good with his hands. He built the air pump with which he and Robert Boyle (1627–91) did many experiments—together they invented the barometer. He greatly improved clock design with the anchor escapement and the clock-spring. He invented the universal joint and the conical pendulum (see page 129). He made his own microscope, and wrote the first popular science best-seller, *Micrographia*.

After Hooke died and Newton became President of the Royal Society, Hooke's portrait mysteriously disappeared, and so did his apparatus and notebooks. Newton was certainly a genius, but he does appear to have gone out of his way to bury Hooke without a trace.

And the first time Newton mentioned a falling apple was to his friend and biographer William Stukeley, after a good lunch in 1727.

17. What is the attraction of mountains?

A weight hung on a piece of string is called a plumb bob. Isaac Newton (1642–1727) described how gravity pulls a plumb bob straight down toward the center of the Earth, but he realized there was sometimes a problem. If you were standing beside a large mountain, the plumb bob would be attracted by the mountain, and therefore would not hang straight down, but would be pulled a little sideways toward the mountain.

Newton called this "the attraction of mountains," and for a hundred years it was regarded as a nuisance by surveyors of mountainous regions. However, in 1770 the Astronomer Royal, Nevil Maskelyne (1732–1811) had a brilliant idea. If you could measure the attraction of a particular mountain and also estimate its mass, then from the relative attractions of the mountain and Earth you could work out the mass of Earth.

Surveyor Charles Mason (1730–87) was dispatched to the highlands of Scotland, for wages of half a guinea a day, to ride around and find the best mountain. His answer was Schiehallion in the Cairngorms (opposite). Maskelyne tried to persuade Mason to go back and do the experiment, but in the end he had to go himself, reluctantly, for he certainly didn't feel the attraction of mountains.

From cozy Greenwich, England, Maskelyne set off by sea to Perth and then 45 miles (70 km) north on horseback into the hills, and set up camp on the south side of the mountain. In dreadful weather he observed the stars overhead, and fixed his latitude using the heavens and a plumb bob. Then he moved around to the north side of the mountain and repeated the process. In all he made 337 astronomical observations, and worked out that the distance between his two camps was 1 mile, 480 yards (2 km).

Meanwhile, a team of surveyors, tramping through the bog and heather with chains and theodolites, measured the actual distance on the ground to be exactly 1 mile (1.6 km). The discrepancy, 480 yards (0.4 km), was the direct result of the "wrong" verticals, caused by the attraction of Schiehallion. The result, after much tedious calculation, was the first ever measurement of the mass of Earth, five million million million tons, which was within 20 percent of today's accepted value of six million million million tons.

5

18. How do they dig tunnels?

In order to tunnel under the River Thames from Rotherhithe to Wapping in the East End of London, England, French engineer Marc Brunel (1769–1849) invented the tunneling shield, apparently inspired by the shipworm, *Teredo navalis,* which bores into wooden ships.

Imagine 12 men standing side by side in narrow cubicles, 12 men below them and 12 more above. Each man has 12 horizontal planks in front of him. He removes one plank, digs out 4 inches (10 cm) of soil, replaces the plank and takes out the next. When all 36 men have dug the soil from behind all the planks, the entire shield is pushed forward 4 inches (10 cm) by jackscrews, and the whole laborious process begins again, while bricklayers line the new section of tunnel behind the shield.

The men had to work in near darkness, with only candles to see what they were doing, and the air was foul, because the tunnel roof leaked continuously, and what trickled through was raw sewage from the polluted river. Marc Brunel became ill; his son Isambard took over, and almost died when the tunnel flooded in January 1828. Marc finally completed the tunnel in 1842, and it is still in use.

Tunnelers today use the same principles but with high technology. The Eurostar railway between London and Paris has tunnels underneath not only the Thames but also the English Channel, and to build them the tunneling shield has given way to the Tunnel Boring Machine, or TBM.

The TBM is a thousand-ton mechanical megaworm, 650 feet (200 m) long, with built-in kitchen and toilet facilities, and at the front huge tungsten teeth that chew their way through the ground. The spoil is pulled back through the machine and taken back on a conveyor belt to the last exit point, which may be several miles away. The megaworm crawls forward, progressing 5 feet (1.5 m) in about half an hour, depending on the type of soil. Then it stops, rests its teeth and lines the new section of tunnel behind it with a ring of 10 precast concrete segments. Concrete foam is pumped into the narrow gap outside the new ring, and the TBM is ready to move again.

This photograph shows a supply train taking more concrete segments from the entrance to the TBM 4 miles (7 km) along the tunnel.

water

19. How do water drops bounce?

Heavy rain bounces off a wet road almost as if the surface were elastic, and to some extent it is. The drops bounce not off the tarmac but off the water that is already there. When a drop of water lands on a dry, solid surface, it just splats out to form a little pool. But a drop bounces well on a water surface. The surface of water behaves as though there is a skin on it, because of surface tension (see page 10), and this is partly why raindrops bounce.

Here the water in a big plastic box is about 1 inch (2.5 cm) deep. When the drop hits the surface the first thing that appears is a depression—a circular dent in the water surrounded by a slight rim made from the water pushed out of the depression. The kinetic energy of the falling drop has been converted into potential energy in that lifted water and the depression—it's like a micro-tsunami.

Then the rim collapses and pushes water both outward and inward. The water moving out forms circular ripples that spread away from the center. The water moving inward collides at the bottom of the depression and throws a drop about half an inch (1 cm) up into the air. For an instant it hangs, poised, on the tip of a thin column of water, and then it falls back to the surface, sending out more ripples.

Challenge: Is the drop that is thrown up the same water as the original drop that falls? Perhaps you can design your own experiment to investigate.

Milk will not behave like this. A drop of milk falling into a saucer of milk will not throw a drop back into the air, because milk is too viscous.

20. Why does a milk drop make a crown?

This is the second drop of milk falling on a piece of black Plexiglas. The first drop just goes splat and forms a little pool, but the second drop falling into this pool kicks up a crown of droplets all the way around. The kinetic energy of the falling drop is transferred to the milk in the pool and throws the droplets into the air. The third drop makes a mess, because of all the splashes from the second.

Milk is a suspension of fatty solids in water, and it behaves rather differently from pure water (see page 46). There are several reasons for this. First, milk is more viscous (stickier) and this prevents it from bouncing quite as freely as water. Drip milk into a glass of milk and it will not bounce as water does. Instead, it forms a blunt crown and the droplets do not fly off.

Second, milk has a slightly different density and surface tension as water, so it is bound to splash differently. A possible third reason is that milk is slightly inhomogeneous; that is, the consistency varies as you go through it millimeter by millimeter. Some bits are almost pure water, some are thick lumps of fatty solid—this is whole milk—and some are somewhere in between.

For one reason or another, I have never managed to photograph a crown that is perfectly symmetrical. One side always seems to flop more than the rest, and the droplets are of varying sizes. Because the falling drop is never perfectly symmetrical, and because it does not land exactly in the center of a pool that is not quite perfectly circular, the droplets that are thrown up will always vary slightly. I have tried it with semi-skimmed and other types of milk, but did not notice much difference.

I don't feel like a complete failure about this, however. The first person to take such pictures of milk drops was an American genius called Harold "Doc" Edgerton (1903–90), who started in the 1920s, became a Professor at Massachusetts Institute of Technology, invented the stroboscope, and took thousands of stunning photographs, from athletes in action to bullets in flight. Edgerton said that he tried for 25 years without ever getting a perfectly symmetrical milk crown.

21. Why does water form drops?

A thin stream of water from a tap or a pipe is affected by three main forces: gravity, drag, or air resistance, and surface tension.

Surface tension is like a pulling force within the surface that always tries to reduce its area. The surface tension within the stream keeps it cylindrical, because that is the stream shape with the minimum surface area.

DID YOU KNOW?

Anything that falls is pulled down continuously by gravity, and therefore accelerates. In a vacuum any object would fall at nearly 32 feet per second (10 m/s) after one second, 64 feet per second (20 m/s) after two seconds, 96 feet per second (30 m/s) after three seconds, and so on.

Because it is accelerating all the time, each little segment of water is being stretched as it falls, and so the stream becomes thinner and thinner. When a stream segment becomes longer than its circumference it becomes unstable, and forms "capillary waves," which you can see in the photograph, until in the end it breaks into short blobs. So when heavy rain looks as if it is falling in streams, which cannot actually be true, because the streams must have broken up and the streaks of rain must be an illusion.

Once the blobs have formed, surface tension reduces their surface area by pulling them into spheres, but as they change shape the blobs tend to oscillate like jelly. Only near the bottom of the picture are the blobs becoming truly spherical.

Challenge: What shape are raindrops?

People often think that water drops are the shape of traditional teardrops, with a tail above them. Wrong. If they were stationary, water drops would be true spheres. When falling through the air, however, they are slightly distorted by the third force, drag. As the drops fall faster, drag becomes more important; when a water drop has fallen more than 33 feet (10 m), the drag is strong enough to flatten the underneath slightly. So if you could take a snapshot of a falling raindrop it would actually be shaped rather like a bun, with a slightly flattened bottom.

22. Who cares about splattering?

When this photograph—a drop of water landing on a wet ceramic tile—turned up in the *New Scientist* magazine (October 9, 2004) I learned that splattering is a hot research area.

Rain on a windshield impairs the vision of the driver; rain on windows makes dribbles of dirt. Researchers have found two potential solutions to these problems.

A French team followed the birds. Feathers are full of air, which is one reason why they don't get wet (see page 18). So the scientists etched the glass surface to make a forest of micro-spikes only 100 nanometers (nm) high (that's $^{1}/_{10,000}$ of a millimeter) and 200 nm apart. To raindrops this looks like a bed of nails, and the drops become "fakir drops," lying on top without sinking in. The water is effectively lying not on glass but on air, and does not wet the surface. So raindrops do not splatter, but simply roll off.

Meanwhile, an American team went in the opposite direction. They coated the glass with a thin film of titanium dioxide—too thin to affect the transparency. Ultraviolet light from the sun ionizes the titanium dioxide, which makes it immensely attractive to water. Again the raindrops do not splatter; instead they spread out instantly to a thin sheet that soon covers the entire glass surface. The water does not impair visibility, and runs off in sheets, carrying away dirt and debris.

The company sells this material as "self-cleaning glass," and hopes to develop it to make eyeglasses that do not need cleaning and a coating for ships' hulls that is impervious to barnacles.

A French company making herbicides and pesticides was concerned because 80 percent of their products bounced off the waxy leaves of the target plants and ended up polluting the soil. Their solution was to prevent splattering by including polymers made from guar—similar to those used to thicken yogurt—and now they claim that 70 percent of the spray stays on the leaves.

Finally, vulcanologists investigating the behavior of molten rock when it spews out of volcanoes in fire-fountains have discovered that it splatters in a way that is mimicked by mixtures of water and golden syrup—which are much easier to study in the laboratory.

23. Why is a shower warmer in the middle?

A hot shower is a wonderful thing—cleansing and refreshing after vigorous exercise, and probably less wasteful than a deep, wallowing bath. It's also a good place to think about the weather. I once played squash with a meteorologist, and in the showers afterward he asked me why the water was warmer on the inside of the spray. I tried, and it was.

Challenge: Test it for yourself. In the middle of the shower the water may be hot, but nearer the edge of the spray it is noticeably cooler.

At first I thought this might be caused by inefficient mixing; if the hot water is piped to the middle part of the shower head and the cold water to the outside, and they never mix properly, then the mystery would be solved. However, most showers get their hot and cold water mixed at, say, chest height and so it must be properly mixed by the time they reach the top.

Can you work out why the water is warmer in the middle? Think before you turn the page over.

The answer (according to the meteorologist) lies in the cooling effect of evaporation. A shower produces thousands of drops of hot water. Some of that water will evaporate, turn to steam and later condense on the walls—which is why the walls of the bathroom are always wet after a shower. Energy is needed to evaporate water and turn it into steam. This energy is called the latent heat. Therefore every bit of evaporation cools the drop from which the water has evaporated, which is why you tend to feel cold after you have had a swim, and why the water is hottest up near the shower head and cooler down near the ground.

But why is it cooler on the outside?

In the middle of the shower stream, water is evaporating from hundreds of drops, and the air between the drops is saturated with water vapor. That means that water is condensing almost as fast as it is evaporating; the net effect is very little evaporation and very little loss of latent heat.

On the outside, however, the air is not saturated, because there is always fresh, drier air near the shower stream. As a result, the water can evaporate faster from the falling drops, and they cool down more quickly. And that is why the shower will always be cooler on the outside.

24. When will the water run out?

We tend to take clean water for granted, forgetting that we have not had it for long and that most of the world's people don't have enough.

The human body is mostly water, and in common with all other living things on the planet it needs water to survive. Water carries food around the body and takes away the waste products. We lose water when we go to the bathroom, and we lose water as vapor every time we breathe out. To replace those losses we have to drink the stuff.

There is plenty of water on Earth—the oceans are full of it—but most of it is contaminated with salt and other things that make it unfit to drink. Purifying it is possible, but not cheap. Seawater can be desalinated. The simplest way is to boil it and then condense the steam. Salt is left behind when the water boils—and so the steam condenses to make pure water. Unfortunately, this is expensive, because a lot of energy is required to boil water (see page 96).

Removing harmful bacteria is usually done chemically, by adding chlorine, which is cheap and kills bacteria. Water in swimming pools is heavily chlorinated, which is why it has a characteristic smell and sometimes irritates the eyes.

In the West we waste water dreadfully. We water our lawns, take large baths or lengthy showers and flush the toilet several times a day. The average American uses 80 gallons (300 L) a day, while the average person in Africa uses only $2^1/_2$ (10 L)—that's about one flush of a Western toilet. Of all the water-supply problems, the deadliest is contamination of drinking water by sewage. Throughout human history this has been one of the most lethal killers, especially of babies.

DID YOU KNOW?

In British cities before 1830 the infant mortality rate was 50 percent—half of all the babies who were born did not survive until their fifth birthdays. The immediate causes of death were cholera, typhoid, dysentery or just plain diarrhea. In many parts of the world the situation is still desperate; in the time it takes you to read this page, eight babies and children will have died because they did not have clean water.

THE CLOSET OF THE CENTURY

25. Where does "spend a penny" come from?

Everyone has a euphemism for when they want to urinate—wee, pee, piddle, widdle, strain the potatoes, water the tomatoes, explore the geography of the house, pluck a rose, take a leak or spend a penny.

In 1851 the Victorians decided to stage a Great Exhibition of the Industry of All Nations in London's Hyde Park. There was a grand competition to design a temporary building to house the Great Exhibition, and designs were submitted by railway engineers Isambard Kingdom Brunel (1806–59) and Robert Stephenson (1803–59), who were also judges.

Both their designs looked rather like railway stations—surprise, surprise—and were quickly thrown out. The unexpected winner was gardener Joseph Paxton (1801–65), with an extraordinary structure of cast iron and glass that was really a giant greenhouse and had the great advantage of being modular—the sections could be made anywhere in the country and simply be bolted together on site, which made it much easier both to build and to dismantle when the Great Exhibition was over. A cynical newspaper jeered at this "Crystal Palace," but people loved the building, and the name stuck.

Flamboyant plumber Josiah George Jennings was hired to install public toilets in the Crystal Palace—the first major installation of public toilets in Britain. When he had finished he announced to the Commissioners that each user of a cubicle would have to pay one penny for the privilege. They were horrified, and said that visitors would be there to see the exhibition, not to go to the bathroom. Nevertheless, countless men used the urinals for free, and 827,000 people chose to spend a penny, which is probably where the expression originated.

The plumbing business took off in the second half of the 19th century, with the advent of piped water and proper sewers. Jennings took out patents in 1852 and 1854 for water-closets that were almost identical to the flush-out and flush-down toilets we use today; few of the many thousands of patents taken out in the following 50 years brought major changes. Like all his rivals, he kept striving for improvements. Never one to be modest, he eventually produced "the closet of the century," which would have had an oak or mahogany seat, and was flushed by pulling up the handle.

26. Which way will a whirlpool whirl?

You often make a little whirlpool in a bath or sink when you pull out the plug and let the water run out. At first a dent appears over the drain, and water runs in from all directions, but then the flow often begins to twist, and a whirlpool develops. There is a theory that predicts which way the water should whirl, but it needs testing.

DID YOU KNOW?

When there is an area of low pressure in the atmosphere—an anticyclone or "Low"—air flows in to compensate, but it doesn't flow in straight, but in a spiral. Because the Earth is spinning, the wind blowing in toward the Low is swept into a whirl, which blows counterclockwise around a Low in the northern hemisphere and clockwise in the southern hemisphere.

The same thing should happen to water. When you pull out the plug you create an area of low pressure—the dent above the drain. If the water begins to swirl, then in principle it should swirl counterclockwise in the northern hemisphere, and clockwise in the southern. However, there are many minor problems that can interfere. You may accidentally give the water a little swirl as you pull out the plug, or the water may have been moving slowly around ever since you filled the basin. There may be a piece of soap stuck in the drain—or the drain itself may be asymmetric and give the escaping water a bias. Therefore a single experiment is inconclusive.

I would love to conduct a mass experiment—perhaps through the newspapers, or via TV or radio, and get a million people each to pull a plug out and note which way the whirlpool goes. Ideally, each person might do it three times; the more whirlpools we have the better our chances of eliminating chance and finding a real effect underneath.

Challenge: The next 10 times you pull out the plug from a sink or bath of water, notice whether a whirlpool forms, and if so, whether it is clockwise or counterclockwise. Write and tell me your result and which country you live in. I will add up the responses, and if I get enough data I will publish the results on my website.

27. How can a paper clip float?

Generally, a solid object will float in a liquid if it is less dense than the liquid.

DID YOU KNOW?

One milliliter (or 1 cubic centimeter) of water weighs 1 gram; so water has a density of 1 gram per milliliter, or 1 g/ml. Therefore, 1 liter of water weighs 1 kilogram, and 1,000 liters of water, or (1 cubic meter), weighs 1 metric ton.

Anything with a density of less than 1 gram per milliliter should float in water, and anything with a higher density should sink. This follows from Archimedes' Principle, named after the brilliant Greek scientist who lived in Syracuse on the island of Sicily in the third century B.C. Archimedes' Principle says that when an object is immersed in water it feels an upthrust equal to the weight of water displaced.

A wine cork has a volume of about 14 milliliters ($^1/_2$ fl. oz.) and a weight of about 3 grams ($^1/_{10}$ oz.). Push it under water and it will feel an upthrust equal to the weight of water displaced. Because its volume is 14 milliliters it will displace 14 milliliters of water, which weighs 14 grams ($^1/_2$ oz.). So the water will push up on the cork with a force of 14 grams, but since it weighs only 3 grams the upthrust is much greater than its weight, and the cork will bob up to the surface. On the other hand a quarter, which weighs about 8 grams ($^1/_4$ oz.), has a volume of only about 1 milliliter ($^1/_{30}$ fl. oz.); so put it under water and it will feel an upthrust of only 1 gram ($^1/_{25}$ oz.), much less than its weight, and so it will sink to the bottom.

Paper clips are made from metal with a density about the same as a quarter; so they will sink if you drop them into water. However, if you lower a paper clip carefully onto the water surface, using a tissue as a sling so that you disturb the surface as little as possible, you can get it to float.

The photograph shows the paper clip pushing the surface down almost like a rubber sheet, because of the phenomenon of "surface tension." The paper clip floats in a little "boat" that is mostly full of air, but if you disturb the surface the paper clip will plunge to the bottom.

Challenge: Try floating a paper clip, using a tissue as a sling, and see how long it will stay afloat.

28. Does oil really calm troubled water?

Benjamin Franklin (1706–90), the American printer, diplomat and scientist, was crossing the Atlantic in 1757 when he noticed that the wakes of two of the ships in the convoy of 96 were much smoother than the others. He asked why this should be so, and the captain told him, "with an air of some little contempt," that the crew must have thrown the remains of their dinner overboard, and the oily residue smoothed the waves.

Franklin was intrigued. He had heard of pouring oil on troubled waters, but this was the first time he had seen it in action, and began to collect stories, for instance, of Mediterranean divers who went down with a mouthful of olive oil and released it in order to smooth the surface and so get better visibility for finding the fish on the bottom, and spear fishermen in Bermuda who used oil on a rough sea to get a better view of their prey.

Franklin took to carrying a sample of oil in the handle of his walking stick and trying it out on any rough water he came across in England—first on the pond in Clapham Common in London and then on Ullswater in the Lake District.

Oil does not mix with water, but floats on the surface and spreads out enormously. The oil layer becomes so thin that you can see rainbow colors in it; these are caused by interference, as the thickness of the oil film becomes comparable with the wavelength of light (see page 86)—although the colors in this photograph are artificial. Eventually, the oil will spread until it is only a single molecule thick. If Franklin had realized this he could have calculated the size of a molecule of oil and founded surface chemistry.

How it calms the troubled water is not clear, but the layer of oil may damp down the little breaking wavelets before they become big waves.

Challenge: Try putting a teaspoonful of oil on a stretch of rough water in a pond, lake or in the ocean. Use only vegetable oil, which is biodegradable. Does the oil have any effect on the surface roughness? Watch from a distance so you can see where the oil is.

29. Why does a match catch fire?

Starting a fire is difficult. I have never succeeded in making a fire by rubbing two sticks together, or striking a flint with iron to make a spark and igniting tinder. I can do it with a magnifying glass and bright sunshine, but the sun does not always shine brightly, especially at night.

The invention of the friction match by John Walker (c. 1781–1859) in Stockton-on-Tees in 1827 made lighting fires a great deal easier; some people claim this was the perfect invention, since at a stroke it completely solved the problem. The story goes that he was experimenting with various chemicals deposited on the end of a stick, and by mistake scraped the end of the stick on the side of his fireplace, producing a burst of flame and a flash of inspiration. He sold his first matches on April 7, 1827, and eventually called them "Friction Lights," but he never bothered to patent his invention.

The modern match is a simple development of John Walker's original discovery. Rub the match-head on any rough surface and the friction makes it hot enough to ignite. Safety matches, invented in 1855, will ignite only when struck on the side of the box because the black strip contains red phosphorus.

Today about 20 chemicals are used in making matches, including the compound phosphorus sesquisulphide P_4S_3. You can tell that sulfur is still involved because when the match is first struck you can smell the sharp aroma of sulfur dioxide in the smoke.

30. How does a candle burn?

Candle wax is a mixture of long-chain hydrocarbons, and like all hydrocarbons, its molecules need to be surrounded by air before they will burn, which means turning them into vapor and mixing the vapor with air.

Gasoline is also made of hydrocarbons, but with much smaller molecules, and so it is much more volatile. Gasoline has a strong smell, boils at about 212°F (100°C), and turns easily to vapor, while solid candle wax has almost no smell and boils at a much higher temperature. Gas will burn when you put a match to it, because there is enough vapor above the liquid at room temperature to ignite, and the heat from the flame then makes more liquid evaporate.

DID YOU KNOW?

Modern candle wicks are clever. They are under more tension at one side, so when the solid melts away they bend over. This ensures that the wick does not stand up too high and make the candle smoke.

Wax does not light so easily. Put a match to solid wax, or even liquid wax, and it will not catch fire. The trick is in the wick, which is a piece of string. The heat from the flame melts the solid wax at the top of the candle to make a pool of liquid wax, which soaks into the wick and is drawn up it by capillary action. When it gets to the top, the liquid wax is boiled, or vaporized, by the heat of the flame, which burns at 900°F (500°C) or more. The vapor then drifts away from the wick and burns in the air being drawn up from below.

The most celebrated person to study candles scientifically was Michael Faraday (1791–1867) at the Royal Institution in London, England. In 1855 he delivered a set of Christmas lectures on the chemical history of a candle.

Challenge: Try one of Faraday's demonstrations. Light a candle and a match. Blow the candle out gently—or even better snuff it out—and then immediately hold the lit match just above the wick. You should see the flame jump back down, showing that what comes off the wick just after the flame goes out is not smoke but wax vapor, with its funny smell.

31. Why are flames bright?

Starting a fire with big bits of wood is difficult. Lighting a wood fire is much easier if you start with twigs or shavings and gradually build up to larger pieces. Small pieces catch fire more easily because they have a large surface area. Imagine slicing a slice of bread to make two ultrathin slices. This will almost double the total surface area, since you now have four faces. The same principle applies to wood: the thinner the slices the greater the surface area. The greater the surface area the more quickly the inside of the wood heats up and releases the vital inflammable gases.

Flames are chemical reactions between inflammable gases and oxygen in the air. Most solids and liquids don't burn, but some of them give off inflammable gases when they are heated. That is how wood burns. When it gets hot it oozes gases that can react with oxygen; if there is a flame nearby, the gases will catch and the wood itself will appear to be burning.

When the fire gets hot enough, large pieces of wood will burn, although most of the reaction is still between gases.

DID YOU KNOW?

Bright flames are dirty; clear flames are clean.

When hydrogen gas burns in pure oxygen, the flame reaches about 3,600°F (2,000°C), but it's almost invisible—a pale blue apology of a flame. The reason is that hydrogen burns to give only one product—H_2O or steam. The steam goes off into the air, and there are no particles to glow.

When a candle burns (see page 69), the wax hydrocarbons burn inefficiently; the hydrogen is stripped off first, leaving behind particles of carbon—soot. Floating about in the flame, these bits of soot are so hot that they glow, and that is what makes the flame bright.

Challenge: Hold a white plate or saucer in a candle flame for a second—not too long, or it may break—and see the soot on it.

These flames from burning wood are bright for the same reason. Wood is mainly cellulose, and the gases that come off are compounds of carbon, hydrogen and oxygen. Once again, hydrogen is consumed first, and the sooty particles left in the flames glow warm red.

32. What are firework sparks?

Gunpowder was invented in China in the ninth century. The Chinese used it for fireworks but soon realized its military potential, and made fire-sticks, cannons and bombs. The Arabs picked up the secrets of the chemistry, probably via the trade routes, and in the 13th century an English monk, Roger Bacon (c. 1214–92), recorded a formula. Worried that it might get into the wrong hands, he wrote it down as a Latin anagram to keep it secret. Bacon's recipe was actually rather feeble—it would only pop and fizz—but perhaps he was again playing safe.

DID YOU KNOW?

Gunpowder is a mixture of 75 percent saltpeter (potassium nitrate), 10 percent sulfur and 15 percent charcoal (carbon). The ingredients should be intimately mixed and ideally ground together into a fine powder if the full power is to be achieved. Be careful! This is a hazardous process, since a spark will cause an explosion.

When gunpowder is ignited, the potassium nitrate releases oxygen, which reacts with the sulfur and the carbon to make the gases sulfur dioxide and carbon dioxide. In most burning reactions the fuel has to mix with oxygen gas, and the mixing process takes time, which means that the burning is relatively slow. In gunpowder the solids are intimately mixed, which means that atoms of oxygen are formed right next to atoms of sulfur and carbon. There is no delay for mixing, and so the reaction is fast.

Light a small heap of gunpowder and it goes FFFT!, burning almost instantly and producing a cloud of sulfurous smoke—but there is no explosion. To make an explosion you have to confine the gunpowder in a container—even simply wrap it in tape; then when ignited the gunpowder explodes. The combustion produces large volumes of gas, and if the gas cannot instantly escape it will burst out of the container, and that is the explosion.

Most fireworks are based on gunpowder, contained in various cunning ways to produce whooshes and fizzes and whistles and bangs. The colors come from metal salts included in the gunpowder mixture—strontium for red, sodium for yellow, copper for blue. The sparks, as here, come from tiny pieces of metal—iron, zinc or aluminum—burning in the air.

33. Why does electricity make sparks?

This picture is a fraud and a fantasy. In real life, sparks cannot come out of a plug like this, because the electrical energy comes from the socket in the wall, rather than from the plug. This picture is merely a parable of how dangerous electricity can be.

Each of the dozens of sparks in this picture is like a tiny flash of lightning, for as Benjamin Franklin (1706–90) showed, a flash of lightning is an electric spark. Air is a good insulator, but electricity will spark through it when the voltage is high enough to break down its resistance.

DID YOU KNOW?

Dry air breaks down at about 30,000 volts per centimeter (1 cm = $^{2}/_{5}$ in.). Suppose you have two pieces of metal, one at 0 volts and one at 30,000 volts; if you hold them less than 1 centimeter apart a spark will jump from one to the other, but if you hold them more than 1 centimeter apart there will be no spark.

The voltage needed is proportional to the distance; so these sparks, about $^{3}/_{4}$ to 1 inch (2 or 3 cm) long, needed a voltage of around 70,000 volts. A lightning strike is caused by a potential difference of many millions of volts, although wet air breaks down more easily than dry air; so lightning, which nearly always comes with rain, needs less than 30,000 volts per centimeter.

The actual mechanism of lightning is complicated. When a huge voltage difference builds up between a cloud and the ground, a "leader stroke" begins to zigzag its way downward, seeking out the path of least resistance and ionizing the air as it goes. It charges up molecules of air by knocking electrons off some of the atoms and on to others. Once the air is ionized it can carry electricity much more easily, so that when the leader stroke reaches the ground it effectively completes a short circuit between cloud and ground, and then the main lightning stroke flashes.

So much energy is dissipated that millions of molecules of air are heated to a very high temperature, and glow brightly. That is why lightning flashes. At the same time the air along the path of the stroke, suddenly white hot, expands explosively, which is what causes the crash of the thunder.

34. Why does your hair stand on end?

Sometimes, when you hear an odd noise in the middle of the night, or when you thought you were all alone and something touches your shoulder, you feel a prickling up the back of your neck and your hair feels as though it is standing on end.

Some animals make themselves look bigger to scare opponents: cats fluff up the fur on their necks and backs. For both humans and cats this is an automatic reflex. The muscles in the skin contract in such a way that all the hairs are pulled up to stand roughly at right angles to the surface.

Much the same happens when you get cold; you can see all the hairs on your arms standing up. Small birds fluff up their feathers in cold weather and look bigger than usual. When the hairs or feathers stand on end they trap the maximum amount of air between them. Air is a good insulator; a layer of air around the skin does a good job of preventing heat loss, which is why loosely weaved undergarments are good for keeping you warm. This air-trapping maneuver works well for birds. Unfortunately, most humans have body hair that is too sparse to trap the air effectively.

For this photograph the hair was charged up to around 100,000 volts. Because every hair has a negative charge, and like charges repel each other, all the hairs push one another as far apart as possible, which makes them all stand on end.

During the 18th century, scientists and inventors made a variety of devices to generate static electricity both for research and for practical use. Electricity was widely thought to be therapeutic, and many doctors used electrical machines to give their patients slight shocks; indeed such therapy may have been used centuries earlier (see page 78). Showmen also developed party tricks—the most spectacular involved the electric monks. In 1746, Abbé Jean-Antoine Nollet persuaded 200 white-robed monks to stand in a circle a mile around, each holding a piece of iron wire in each hand to connect him to the monk on either side. When Nollet completed the electrical circuit with a charged capacitor called a Leyden jar, the entire mile of monks jumped simultaneously into the air.

35. Who made the first electric battery?

In 1799, the Italian scientist Alessandro Volta (1745–1827) discovered that when he put two different metals together—copper and zinc—they generated electricity. This was an exciting find, and news spread fast. Volta wrote to Joseph Banks (1744–1820), President of the Royal Society in London, and soon many people in Britain were investigating.

They constructed various types of cell to try to make the process more efficient; they stacked the cells to make a "battery." They used the electric current in a variety of ways: Humphry Davy (1778–1829) isolated sodium metal, Michael Faraday (1791–1867) built the first electric motor, and so on.

Volta may not have been the first, however. This photograph shows the "Baghdad battery." It was discovered near Baghdad in an archaeological site that dates back some 2,000 years to the ancient Mesopotamians. Other similar examples have been found since.

There is an earthenware pot, a copper cylinder about 6 inches (15 cm) long and an iron rod that fits inside the cylinder but seems to have been separated from it by a plug of bitumen—an electrical insulator. How might it have worked? Put the copper cylinder in the pot to keep it upright. Fill it with acid, such as lemon juice, grape juice or vinegar. Insert the iron rod in the cylinder so that the metals do not touch, and across the metals you would find half a volt of electrical potential. Cynics say these things were ornaments or scroll-holders, but I am convinced those Mesopotamians were making electricity many hundreds of years before Volta.

What were the Mesopotamians using it for? They could not have had electric lights—the arc lamp needs a huge current and the light bulb needs high technology. They could not have had motors or telephones. There are two main theories.

First, doctors might have used electricity to treat patients—"Swallow this potion, then hold these metal rods. Do you feel a tingle? Good! The treatment is working; please pay at the desk."

Second, gold will dissolve in chemicals involved in tanning leather. A small silver figurine placed in such a solution could be electroplated with a thin layer of gold—and then it would appear to be a gold figurine, making it much more valuable.

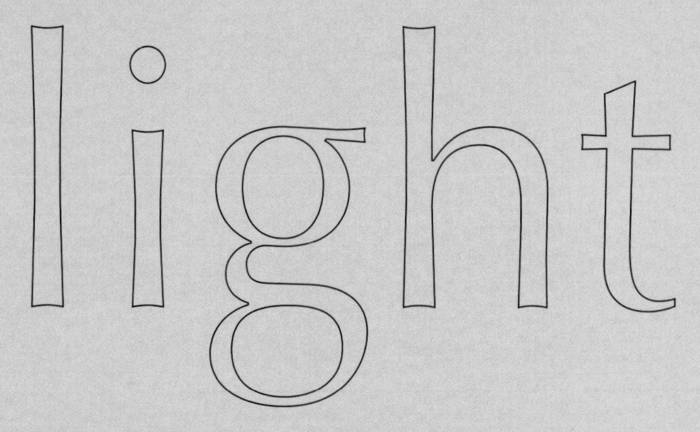

36. What is limelight?

Goldsworthy Gurney (1793–1875) was an enterprising Englishman, who built himself a castle on a sand dune in the seaside town of Bude, Cornwall. He was a champion of steam carriages, and took over the ventilation at the House of Commons—the task of releasing all that hot air.

His steam-carriage business looked promising when in 1829 he was invited to organize a trial mail run from London to Bath. Despite various misadventures they occasionally reached the astonishing speed of 15 miles per hour (25 km/h), and took only half the time of the regular horse-drawn service. Sadly, however, the government decided the future lay with trains—railways were subsidized, heavy tolls were levied on steam carriages and the business failed.

Nothing daunted, Gurney embarked on a new career, as a lighting engineer. He had already invented a blowpipe for burning hydrogen in oxygen, and won a gold medal from the Society of Arts, Manufactures and Commerce. The oxy-hydrogen flame is extremely hot—around 3,600°F (2,000°C)—and he investigated its effect on various things, including limestone.

Limestone is calcium carbonate, but blackboard chalk (calcium sulfate) works equally well. Anything will glow when heated to 3,600°F (2,000°C), but calcium compounds emit a particularly bright white light. As the lump of chalk gets hotter, the light gets brighter and brighter—it has hardly started in this photograph—until you cannot look at it with comfort. Gurney called it limelight.

37. Why does the ruler look bent?

Light travels in straight lines. This fact was demonstrated about a thousand years ago by an Arab scientist named Ibn al-Haytham, or Alhazan (c. 965–c. 1040). If light didn't travel in straight lines we would have no shadows, and we would be able to see around corners, just as we can hear around corners.

However, light can be refracted, or bent, when it passes from one material into another—from air into glass, for example. When light goes through a triangular prism (see page 85), it is refracted through a considerable angle. Furthermore, because of the shape of the prism the various colors are refracted by different amounts, which is why the prism splits white light into its separate colors.

So when light passes from air into water or from water into air, it changes direction at the surface. When you look at this picture, the ruler looks bent, but in fact the ruler is straight and the light rays are bent—or at least those that have come out of the water have changed direction as they emerged. This is hard for the brain to accept; because we are used to seeing light travel in straight lines we interpret what we see as a kink in the ruler.

For the same reason, fish looking up through the surface of a pond see a wide-angle view of the world. Objects just above the surface of the water look higher up and nearer the center of the view than they really are. You can get the same effect looking up from below the surface of a swimming pool, or even a bath if you run it deep enough.

Challenge: Try taking a picture with an underwater camera pointing upward from just below the surface, and then another from just above the surface. Does the water widen the angle of the picture?

We use the refraction of light by lenses to magnify or minimize objects.

DID YOU KNOW?

A magnifying glass or convex lens is a circular prism. Rays that pass into the lens near the middle go straight through, but those that arrive near the edge are refracted inward, which makes the object behind the lens appear bigger than it really is.

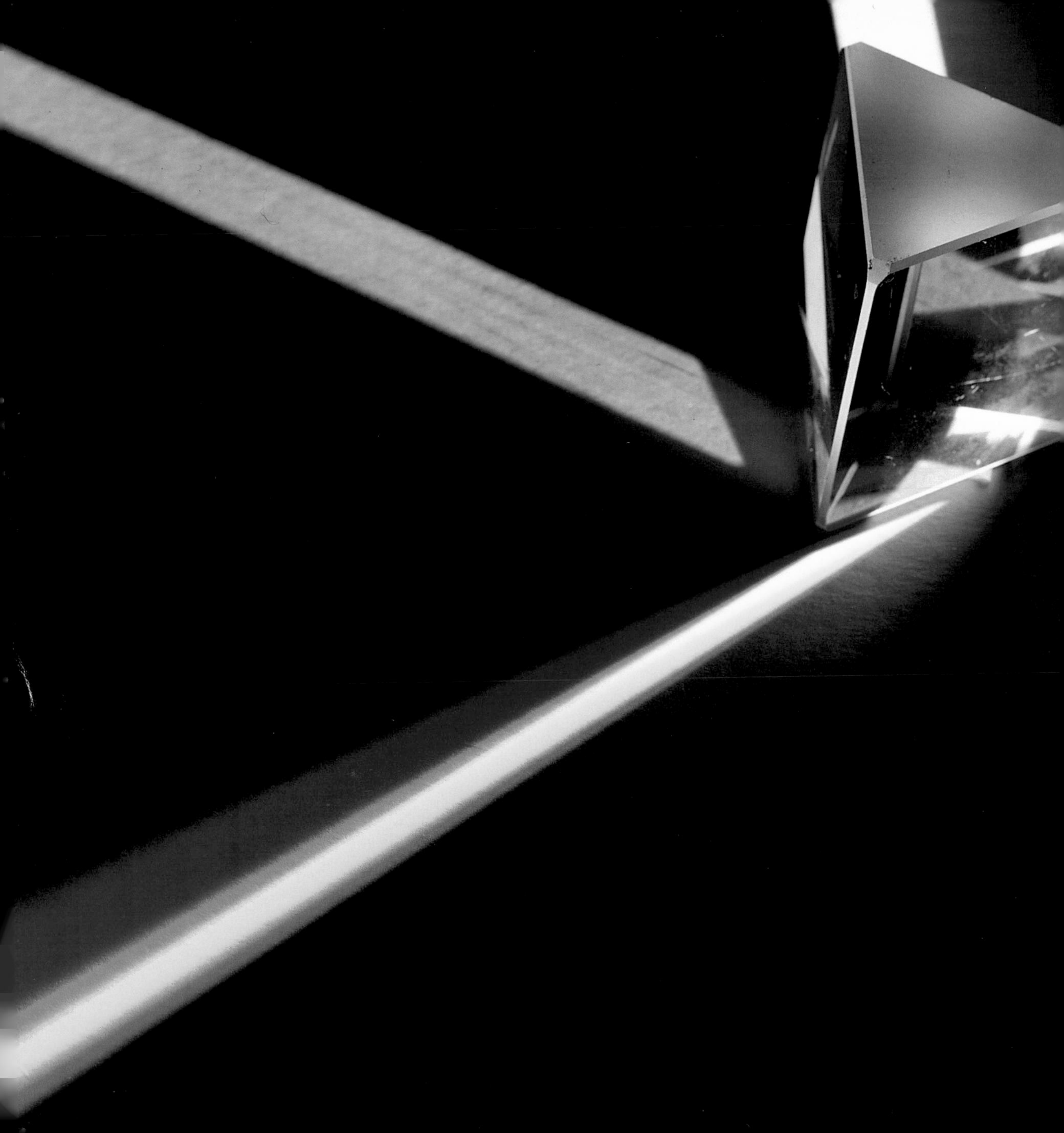

38. Why are rainbows colored?

Light travels in waves a bit like water waves, but much smaller. Waves in the sea may be 30 feet (10 m) from one crest to the next; this distance is called the wavelength. Light waves have much shorter wavelengths: they range from 400 to 800 nanometers (1 nm = $^{1}/_{1,000,000}$ of a millimeter). So even the longest visible light wave is less than $^{1}/_{1,000}$ of a millimeter from crest to crest.

We see light of different wavelengths as different colors. Violet and blue light have the shortest wavelengths; red the longest.

In his first published paper, Isaac Newton (1642–1727) wrote: "In the beginning of the year 1666 I procured me a triangular glass-prisme, to try therewith the celebrated phenomena of colours." He concluded from his observations of the spectrum that white light from the sun is a mixture of all the colors of the rainbow. When light goes through a glass prism it is bent, or "refracted," and the colors are separated because they are refracted through different angles, red the least and blue the most. Newton reported seeing seven colors: red, orange, yellow, green, blue, indigo and violet.

Rainbows are formed in the same sort of way, as sunlight is refracted when it passes through raindrops, once as it goes into a drop, and again as it comes out.

Many people don't see indigo and violet as colors distinct from blue, but on the other hand bees and some other creatures can see beyond the violet into the ultraviolet.

In 1800 William Herschel (1738–1822), a German musician and astronomer, set up a prism to split up sunlight, and placed a thermometer in each color in turn to investigate whether they corresponded to different temperatures. After doing red, he went away to lunch, thinking the experiment was over. When he came back he was amazed to find a high temperature was still showing, even though the sun had moved, and the thermometer was well out of the visible spectrum. Because it was beyond the red, he named this part of the spectrum "infrared" (below the red). Likewise the light beyond the violet is called "ultraviolet."

DID YOU KNOW?

A good trick for remembering the colors of the rainbow is by the first letters of the words in the sentence Richard Of York Gave Battle In Vain.

39. Why are bubbles colored?

The colors produced by prisms and rainbows are caused by refraction (see page 85), but the colors in bubbles and oil patches are formed by interference.

The skin of a bubble is made of a thin film of water and soap mixture. As the liquid gradually evaporates or runs away, the film gets thinner, until it is only about as thick as the wavelength of light, which ranges from 400 to 800 nanometers (nm).

At this stage, the light waves bouncing to and fro inside the film interfere with one another: if two crests come together they form a bigger crest, while if a crest meets a trough they cancel one another out.

Imagine two rays of light with a wavelength of 600 nm reaching a soap film that is just 600 nm thick. Suppose one ray goes right through, while the other goes into the film, is reflected back from the far side, and then reflected again so that it finally emerges alongside the first ray but after two internal reflections. Because their wavelength is the same as the thickness of the film, the crests of the two rays will still be in step and the color will be bright. However, if two rays with wavelengths of only 500 nm arrive, then after two reflections the second ray will be out of step with the first, the waves will tend to cancel one another out and the color will disappear.

The various colors correspond to light of different wavelengths; blue is about 400–450 nm, yellow 600 nm and red 700–800 nm. As a result, interference happens at different thicknesses for different colors; so the bubbles develop bands of color that constantly shift as the bubble film gets thinner. In the end, when the thickness falls below 400 nm, the color disappears altogether just before the bubble bursts.

Challenge: Bubbles thin because of gravity, as well as evaporation, which means that they are thinnest at the top. So you should expect to see horizontal colored bands moving down the bubble, in the order of the colors of the rainbow—red, orange, yellow, green and blue. Before it goes black the last band you see should be blue—why not try it and see?

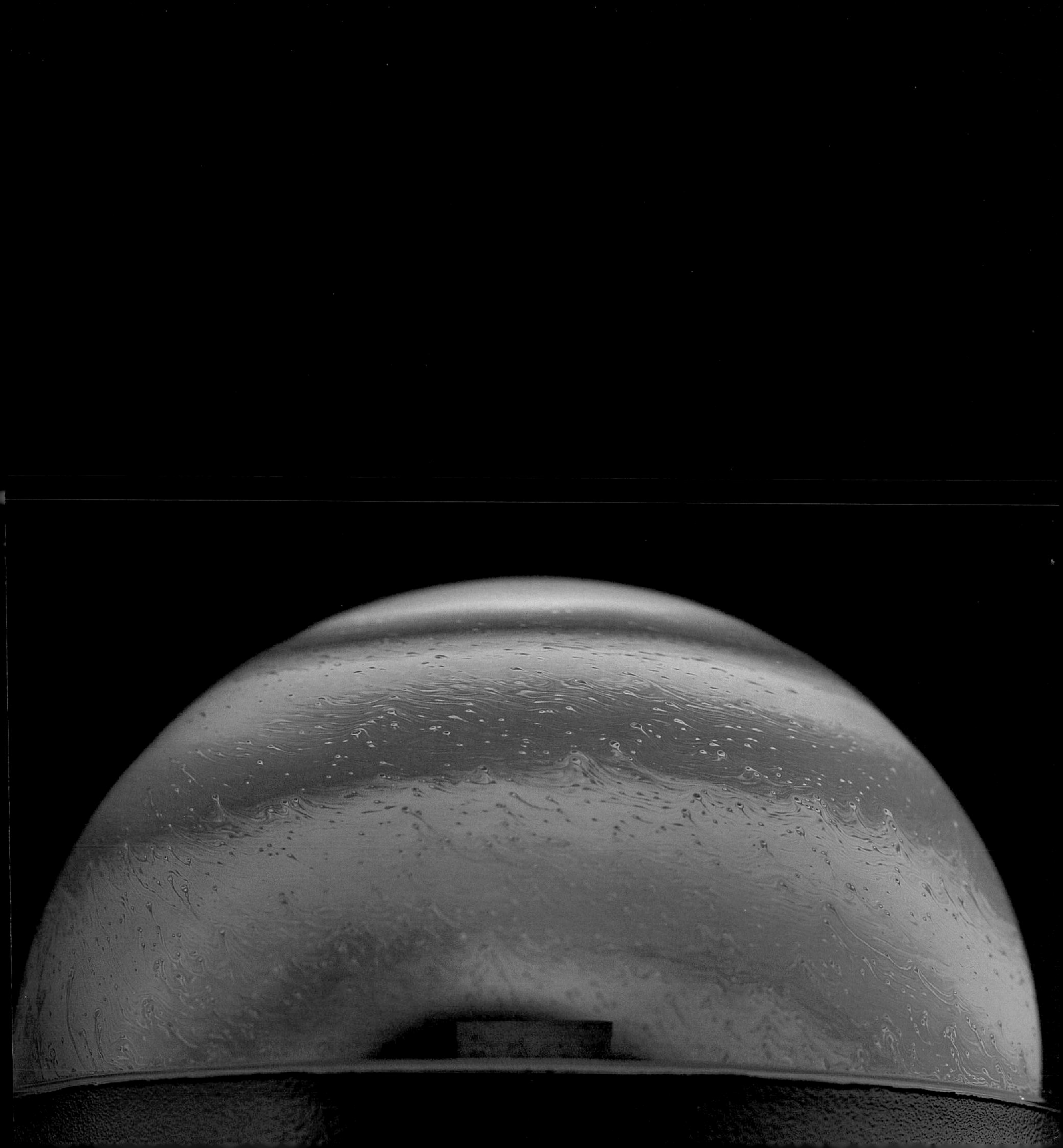

40. Why is the sky blue?

This question was first answered by John Tyndall (1820–93), who in 1867 took over the running of London's Royal Institution from Michael Faraday (1791–1867). A keen mountaineer, Tyndall was one of the first people to climb the Matterhorn. Before he attempted this, he calculated that the energy needed for the climb would be exactly the same as he would get by eating a ham sandwich; so that was all the food he took.

Back at the Royal Institution, he filled a glass tank with filtered, dust-free air, and noticed that he could not see a beam of light passing through it. He could see a beam of sunlight slanting across the room, and he realized it was visible only because the light was scattered by dust particles in the air. In his filtered air there was no dust to scatter the light.

He experimented by filling another tank with water and adding a teaspoonful of small particles (milk powder works well). A beam of light shone through it appeared bluish—the small particles scatter blue light. With a leap of imagination he realized that the sky is blue because small particles in the atmosphere—actually molecules of air—scatter blue light from the sun.

The sun lights up the whole of the sky during the day, and because air molecules scatter blue light in all directions, the whole sky looks blue. The light from the sun looks yellowish because although it is really white, some of the blue has been scattered away, and white minus blue makes yellow.

DID YOU KNOW?

Tyndall suspected that the bacteria and yeasts that make food go bad float on dust particles in the atmosphere. He placed specimens of freshly cooked food in test tubes filled with carefully filtered air and plugged the tubes with cotton batting. Those test tubes are still on display in the Royal Institution, although I would not care to eat the contents today.

At sunrise, as in this photograph, the sunlight arrives at a low angle and has to travel through much more atmosphere than later in the day. As a result more blue light is scattered, the sky looks deeper blue, and the sun looks yellow, orange or even red. At the same time it lights the clouds pink or orange.

41. How do TV shows travel down a cable?

These fiber optics or optical fibers are strands of pure glass only about $^1/_{10}$ of a millimeter thick—thinner than a human hair. Cable television is delivered through fiber optics like these, which are also used for closed circuit television, telephone calls and many other message systems. The information is carried by pulses of light.

The basic idea of shining light down tubes was demonstrated 150 years ago by John Tyndall (1820–93, see page 89). By using a flashlight, he showed that light shining along inside a stream of water would be trapped inside it, even when the stream curved. He called this a light pipe. Light normally travels in straight lines, and you might expect it to escape from the light pipe at a bend, but as long as the bend is not too sharp the rays of light are reflected back off the inside of the pipe in a process called total internal reflection, and no light escapes.

The lasers and other technology of the 1960s allowed this idea to be extended to glass fibers. A beam of light shining into one end of a glass fiber is trapped inside, and comes out of the other end with almost no loss on the way, even though the fiber bends around corners. By sending pulses of light along the fiber you can send any sort of digital information. The fibers are actually made with an outside layer of different glass, and the internal reflections happen at the boundary between the main fiber and the cladding.

DID YOU KNOW?

Optical fibers with tiny cameras are called endoscopes, and are used by doctors for examining the inside of the body and by keyhole surgeons for carrying out operations remotely.

By 1980 fiber optics became the most popular method of transmitting information, since they have several advantages over sending pulses of electricity along a cable. There is far less loss of energy, and signals can be sent 20 miles (32 km) along a glass fiber without any boosting of the signal. What's more, they are not affected by moderate heat or by magnetic fields, they can't make sparks, and are therefore safer in the vicinity of explosives or fuel; and they can't easily be tapped, so they are much more secure.

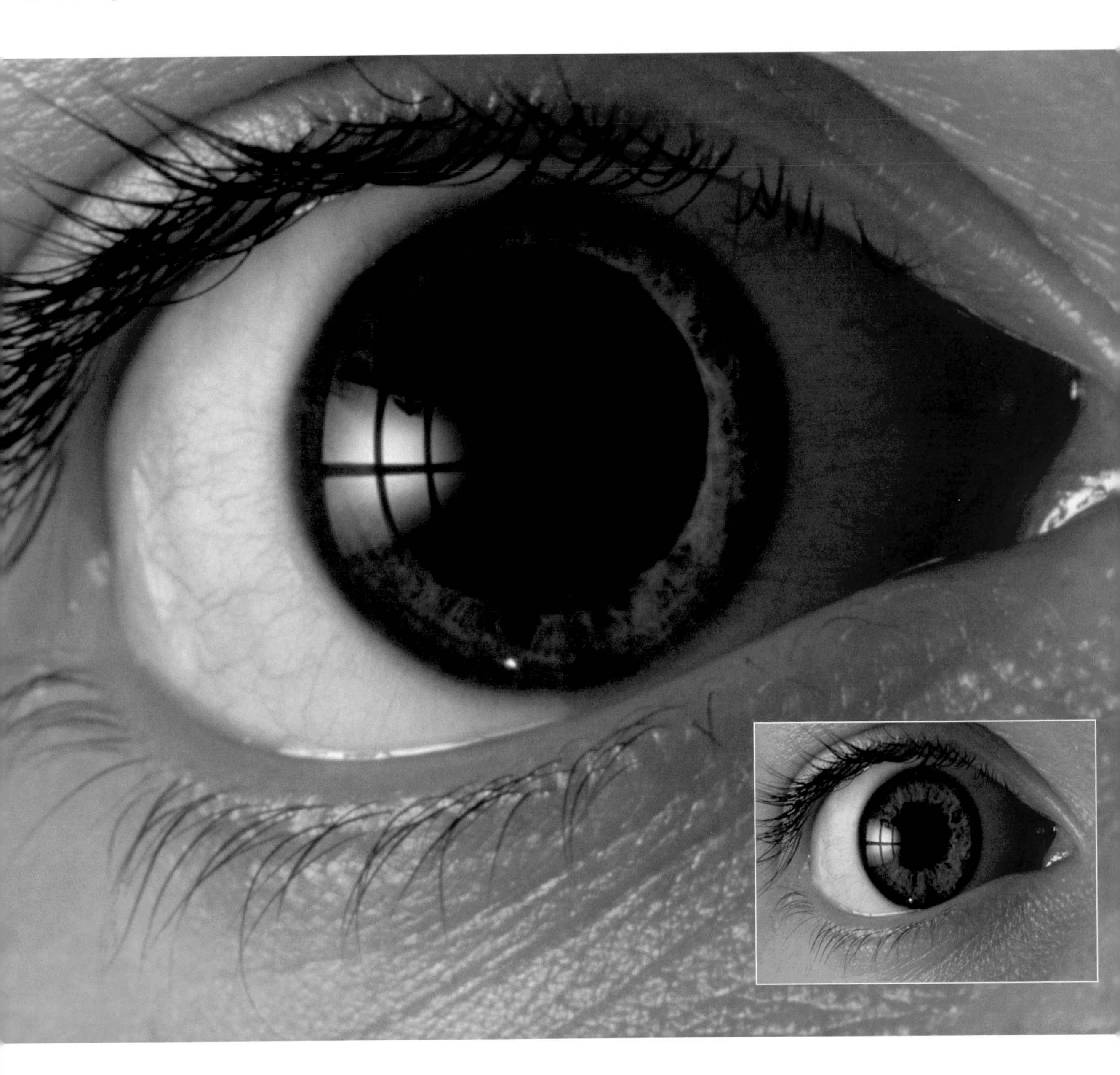

42. Why do pupils dilate?

The first person to work out how we see was the Arab scientist Ibn al-Haytham, or Alhazan (c. 965–c. 1040), around 1,000 years ago. Before he came along people thought that when you look at something, rays go out from your eyes and bounce back from the object. He realized that light from the sun or a lamp is shining anyway, whether you are looking or not. When you open your eyes the light streams in, and allows you to see whatever the light has bounced off. In other words, seeing is a passive process, at least until the light reaches the eye.

Light varies enormously in brightness. The sun is so bright that you will seriously damage your eyes if you look at it; it's a thousand times brighter than a sunlit landscape, which is a thousand times brighter than a moonlit landscape. The retinas at the back of your eyes have only a limited range of sensitivity, and so to help you cope with this vast range of brightness you can vary the aperture in front.

The black pupil in the center is a hole to let the light into your eye. In bright light the pupil contracts and does not let much light through. In dim light the pupil gets bigger, or dilates, to let more light in and give you a chance of seeing clearly even in near-darkness. Your pupils also change size when you are thinking hard, which may be why poker players often wear shades.

Challenge: Try going first from a dark room into bright sunlight, wait until you can see clearly enough to read small print, and then go back again. Measure the time it takes for your eyes to adjust. Is it the same both ways?

When you go from a dark place into a bright one, your eyes adjust quickly, and although you feel an initial shock you can soon see normally. However, most people's eyes react more slowly in the opposite direction; go from bright sunlight into a dark room and you may need several minutes before you can see clearly, as your pupils open up to let in as much light as possible.

DID YOU KNOW?

Your eyes are linked: shine a bright light into one eye and both pupils will contract at the same time.

43. Can shadows form in space?

When I first looked at the full moon through a telescope I found it rather dull and featureless, because the full moon is lit flat-on, from behind me. The half moon below looks much more interesting because it is lit from the side, and near the left-hand edge the craters—and some mountains—are made clear by their shadows, especially at the bottom of the picture.

Early observers thought the large dark areas looked like oceans, and gave each a Latin name. The *Mare Serenitatis* (Sea of Serenity), top right, overlaps the large and irregular *Mare Tranquillitatis* (Sea of Tranquillity), which is where Neil Armstrong and Buzz Aldrin landed from their *Apollo* spacecraft in 1969.

DID YOU KNOW?

Both Galileo in 1610 and Robert Hooke in 1664 drew pictures like this photograph and used them to speculate about the moon's surface.

The craters on the left are revealed by "local" shadows, cast by the moon's own lumps and bumps, but what of shadows thrown by one celestial object on another? The great Portuguese/Spanish explorer Ferdinand Magellan (c.1480–1521) wrote in the early 1500s: "The church says the earth is flat, but I know that it is round, for I have seen the shadow on the moon, and I have more faith in a shadow than in the church." The ancient Greeks before him knew that Earth is a sphere, partly because of the shape of its shadow.

When the moon, in its orbit around Earth, passes through Earth's shadow, it is said to be eclipsed. The multiple exposure photograph above shows a total eclipse of the moon. The images start bottom left, and were taken roughly every half hour until clouds obscured the view.

When you look at these images you can see that a shadow is moving across the face of the moon, and that the shadow is curved. The curvature is not blindingly obvious, however, and since neither Magellan nor the ancient Greeks had telescopes, which had not yet been invented, they must have had sharp eyes and confidence in their sight to assert that Earth must be a sphere.

To begin with, the shadowed part of the moon is merely dark, compared with the brightness of the rest, but then as the shadow grows it seems to turn red. This is because the sunlight reaching the moon is shining just past the edge of Earth and is therefore passing obliquely through the atmosphere twice—first to get to the surface and second to get out again. Therefore an observer on the moon would see, in effect, a double-strength sunset, with much of the blue light scattered away by the molecules in the Earth's atmosphere (see page 89).

Challenge: Have a look at Venus with a telescope or binoculars and explain its shape.

ice & rain

44. Why does ice cool your drink?

This seems a silly question: obviously ice cools your drink because it's cold. However, that is not the whole story. More important is the latent heat.

The ice cubes in this beaker are melting, and the temperature of melting ice is 32°F (0°C). Ice can be colder that this—it may be at 14°F (-10°C) in your freezer—and water can be warmer, but when ice and pure water are mixed, the temperature is 32°F (0°C). So while there is ice in your drink, the whole drink will stay at 32°F (0°C).

In solid ice, the H_2O (water) molecules are tightly bound together. They have to be broken apart to melt the ice, and this takes energy—the latent heat of melting. While the ice is melting, it extracts this energy from the liquid, and so prevents the drink from warming up.

Similarly, when water evaporates, the molecules of liquid need energy to escape into the atmosphere. That energy is the latent heat of evaporation. Heat a pan of water on the stove, and the water temperature will rise steadily, until it reaches 212°F (100°C). Then the water boils, and however much heat you put in, the temperature will not change until all the water has boiled away.

Latent heat was discovered c. 1760 by Scottish chemist Joseph Black (1728–99). Many of his pupils at Edinburgh University were the sons of whisky distillers, who wanted to know why distilling whisky was so expensive. Black realized that to distill whisky—or anything else—you have to give the molecules enough energy to escape from the liquid into the vapor. This energy does not raise the temperature, and so he called it "latent" heat, meaning hidden heat.

-40...120°C
00.0
off
on
Brannan

45. Why do ice cubes crack in your drink?

When most pure liquids freeze, they form solids that are denser than the liquid; so the solid forms at the bottom of the container. Water is almost unique, because ice is about $^1/_{10}$ less dense than water, which means that ice cubes float in drinks and icebergs float in the sea, with about $^1/_{10}$—the tip of the iceberg—showing above the surface.

Because the dissolved salt increases the density, saltwater is more buoyant than freshwater. This means that both swimmers and lumps of ice float higher in the sea.

Water has another curious property; it has a maximum density at 39°F (4°C). A cup (250 mL) of water at 39°F (4°C) weighs more than a cup of water at 37°F or 41°F (3°C or 5°C), and noticeably more than a cup of ice at 32°F (0°C) or below. So when water cools below 39°F (4°C) it expands, becomes less dense, and rises to the surface. Then ice forms on the surface and floats. That is why ponds freeze on top, while the water underneath stays liquid unless the freeze is severe and persistent.

When you make ice cubes in a freezer, the heat is extracted from all around; so each packet of water freezes from the outside. You can often see, before the cube has frozen completely, that there is liquid water in the middle. When the water in the middle freezes it wants to expand—because ice is less dense than water—but is locked in by the cage of ice already formed. So the middle freezes under compression, which puts terrific strain on the cage.

When the ice cube is dropped into a drink, the cage of ice is weakened as it begins to melt or dissolve. Then the pressure from within may break it open; you can often hear it crack and see the cracks in the outer cage, and sometimes it even falls apart.

DID YOU KNOW?

If ponds froze from the bottom, all life in them would be extinguished almost every winter, and it is unlikely that complex life could ever have developed on Earth.

Challenge: When you put an ice cube into a drink you can see the ice is clear around the outside and cloudy in the middle. What can you say about how the water froze? There are clues in icicles (see page 100).

46. Why do some icicles have bubbles up the middle?

DID YOU KNOW?

Some icicles have, in the middle, a column of tiny bubbles perhaps a millimeter in diameter.

The snow on the roof is slowly being melted either by the sun or by heat coming up from the building, so that water trickles down to the edge of the roof. When it meets the cold air, a drop of water freezes and forms a blob of ice. Along comes another drop of water, which freezes onto that blob to make it bigger, and so an icicle begins to grow. Although the snow on the roof is warmed enough to melt slowly, the growing icicle remains below freezing point partly because it is clear, so it does not absorb the sunlight, and partly because it is fully exposed to the cold air around it.

As long as the snow is melting and the air temperature is below the freezing point the icicle will keep growing, even though water drips off the end, because some of the trickling water will freeze onto the cold icicle and make it bigger.

Challenge: Icicles often develop horizontal ribs, like ripples across the flow of water. Can you explain why? Are these ribs like ripples in sand, as on page 144?

How do these bubbles get there? There's a clue in every ice cube (see page 99). The water trickling down contains a small amount of dissolved air, which escapes when the water freezes, since the molecules of air cannot fit tidily into the crystal structure of the ice.

Where the icicle is thick, some of the trickling water freezes onto the surface. The solid front moves outward from the central ice core, driving the air before it, and the air will be lost from the outside.

A drop of water hanging from the tip of the icicle, however, has to freeze from the outside, since there is no ice in the middle. Therefore, a skin of ice forms on the surface, and the solid front moves inward, sweeping the air toward the center. The last bit of water to freeze must be the central core. This is where the air is expelled from the water as it freezes, and this may explain why there is a central column of air bubbles.

Or have you a better explanation?

47. What's the difference between hail, snow and frost?

Hailstones form, often in thunderclouds, when raindrops are swept up to where the atmosphere is so cold that they freeze. This can happen even in the heat of summer, because the air is always cold at high altitudes. Hailstones can get bigger by not falling but "yo-yoing." They fall far enough to get wet from more raindrops and are then lifted by strong updrafts to where the water freezes onto the surface, and then fall again. So hailstones grow in layers, like onions, as more and more liquid water freezes on the outside.

DID YOU KNOW?

The world's biggest recorded hailstones were the size of grapefruit, about 5 inches (13 cm) in diameter, and weighed up to 2 pounds (1 kg).

Both frost and snow form when water vapor freezes directly to solid ice without becoming a liquid on the way. When water vapor freezes slowly in the air it forms snowflakes, and the more slowly it forms the bigger the flakes will grow. Eventually they fall from the sky and settle on the top of branches and leaves, and much more piles up on flat surfaces like leaves than on points or thin twigs.

During winter nights, the ground and plants may cool faster than the air. If their temperature falls below 32°F (0°C) and the air is damp, then molecules of water vapor will freeze directly onto solid objects and form frost. The frost grows first on the most exposed parts of the surface, and on irregularities—lumps, edges and especially points. This is why the prickles and edges of holly leaves get coated with ice crystals, although there is little frost on the smooth flat surfaces of the leaves, or on the berries. Frost also rimes twigs and the edges of blades of grass.

On windows, frost often settles in swirly patterns. There are no sharp points or edges on the clean glass surface; so the first few ice crystals will form anywhere—but then they act as nuclei for other water molecules to freeze against. This is why the frost seems to have grown into fronds like ferns or bracken; it does indeed grow from those first crystals.

Challenge: If water and ice are clear, why do snow and frost look white?

48. How do you make snow?

For most people, snow is a cold, wet nuisance, and they buy brooms, shovels, blowers and other devices to get rid of it. For those who run ski resorts, however, snow is an important asset, and the more snow there is the further they can extend their ski business, both geographically—down the mountains—and in time—into the early winter and late spring.

For this reason cunning technology has been deployed to invent snow-making machines, permanently mounted in strategic positions on the lower slopes. Whenever the snow cover approaches the minimum depth for good skiing they are set to switch on automatically in the early morning and run until the air temperature rises to about 23°F (–5°C). Above this temperature they don't work very well. Unfortunately, the temperature sensors are on the machines themselves, so when they are warmed by the sun, they tend to switch off perhaps half an hour after the sun hits them, regardless of the temperature of the air.

The snow machines work by spraying a fine mist of water into a blast of air from an air compressor or a big fan. Fans are less noisy, but big and cumbersome. The water spray is blown high above the ground and rapidly cooled both by the air around it and by the evaporation induced by flying through the air.

DID YOU KNOW?

Snowflakes are formed naturally in the atmosphere by the direct freezing of water vapor into feathery crystals of ice, without going through the stage of liquid water. Freezing of liquid water drops makes hail (see page 103).

The snow machines technically may be making fine hail rather than snow, but it looks, feels and behaves like snow on the slopes, although in practice it is a bit more dense than natural snow, and lasts a bit longer in warm weather.

Snow-making machines often use tiny particles to seed the ice crystals; water will freeze more easily on such particles. This process is called nucleation, and it allows snow-making to work up to a temperature of about 28°F (–2°C). Silver chloride dust will act as a nucleator, and a commercial one is made from a protein that occurs naturally in grass, trees and other plants, and decomposes rapidly in the soil.

49. What is morning dew?

Sometimes on cool mornings the grass looks gray. Outside surfaces are covered with tiny droplets of water. The spiders' webs glisten, and every leaf is a plate of dewdrops.

Dew is just water, condensed directly from the atmosphere. There is always water vapor in the air, and the amount—the humidity—depends on the temperature, and on how wet the ground is. During rain and fog, the humidity is close to 100 percent; the atmosphere is loaded with as much water as it can take.

On a sunny afternoon the sun warms both the ground and the atmosphere. This increases the concentration of water vapor—hot air can carry more water vapor than cold air. If the temperature drops at night, the air may become supersaturated; it may have more water vapor than it can carry, and some will begin to condense—turning back into droplets of liquid water.

When this happens high above the ground, the droplets form clouds, and if they are concentrated enough they may come together into raindrops.

When the condensation happens near the ground, the droplets form early morning mist or fog, which is often "burned off" when the sun warms it enough to turn the droplets back into vapor.

DID YOU KNOW?

The clouds of white or gray "steam" that you can see coming from a boiling kettle or pan are really made of millions of tiny drops of liquid water. Steam is water vapor, which is invisible.

Sometimes, however, during the night the ground cools more quickly than the atmosphere. Then the water vapor condenses directly onto leaves, twigs and rocks, and millions of tiny droplets of liquid water are formed on their surfaces.

Rain splatters indiscriminately, just making everything wet, but dew appears as tiny separate droplets that may not run together, and do not completely wet the surface. Dew is usually purer than rainwater. As they fall, raindrops sweep up not only polluting gases from the atmosphere, but also other garbage (see page 113). So rain may be quite a cocktail. Dew, by contrast, is water that has condensed directly from the air; so it cannot pick up any dirt on the way, and is as pure as the surface it condenses on.

50. What inspired the cloud chamber?

Ben Nevis, the highest mountain in Britain at 4,406 feet (1,344 m), is the remnant of a 400-million-year-old volcano, most of which has been eroded away by glaciers and soft Scottish rain. Walking up is tedious, because you start almost at sea level, the path is unrelentingly rocky, like a 4-mile (6 km) uneven staircase, and there are lots of other people going both ways.

On top of the mountain is a near-flat plateau about 300 feet (100 m) square, with various cairns and the trig point at one end, and a sheer cliff dropping away to the north, which in bad weather is hard to spot and has claimed a number of lives.

In 1883 the Scottish Meteorological Society constructed a path up the mountain, and on the plateau they built an observatory for studying the weather; details of wind speed, temperature and so on were recorded every hour for 21 years. There was also a hotel, catering presumably to walkers, climbers and meteorologists. When the weather was bad, supplies could be hauled up the mountain by cable. The remains of the observatory, the hotel and the cable lift can be seen to this day.

In September 1894, one of the scientists who worked there, C.T.R. Wilson (1869–1959), was fascinated by the "glory" or "Brocken Spectre"—a faint rainbow around his own shadow thrown by the sun on the cloud below him. Determined to study this further, he went back to Cambridge and built a cloud chamber. This was a glass box that he could fill with damp air. By suddenly reducing the pressure inside, he could lower the temperature and cause the water vapor to condense into clouds. This ingenious device turned out to be the best available detector for such charged atomic particles as electrons and protons, which were being generated by colleagues studying radioactivity and nuclear reactions. As a result, Wilson shared the Nobel Prize for Physics in 1927.

DID YOU KNOW?

Some of the best drugs for treating irregular heartbeat are derivatives of foxglove, *Digitalis purpurea*. The effectiveness of the powdered dried leaves was first discovered by a Birmingham physician, William Withering (1741–99). When he was dying, his friends said "The flower of physic is withering," and his memorial in Edgbaston Old Church has a foxglove carved in stone.

51. Why can't they predict the weather?

Hold a spoon in the stream of water from a tap or faucet and in the water flow you can develop beautiful patterns that vary all the time. Indeed, if you were to clamp a spoon in a fixed position under a tap and take a hundred photographs, each would be unique, for the pattern of water flow changes continuously and unpredictably. This is chaos in action.

Although in principle the flow of water can be calculated mathematically, in practice the outflow from the tap varies slightly all the time, and the tiny variations are enough to cause large and obvious differences in the pattern of flow from the spoon. You can also see chaotic flow in the stream from the tap and in the spiraling patterns of dribbling syrup.

The underlying ideas had been known for centuries, but chaos came forcibly to the attention of the public as well as the scientific community in the 1960s, when meteorologist Ed Lorenz, trying to predict the weather with his computer at Massachusetts Institute of Technology, discovered that he could not get it to make the same predictions twice in a row, because the slightest variation in today's conditions could lead to immense differences in two or three weeks' time. He called this "the butterfly effect," because, he said, the flap of a butterfly's wing over a Pacific island may eventually cause a hurricane on the other side of the world.

One direct result of the chaos in the mathematics of the atmosphere is that long-term weather prediction is not merely difficult but actually impossible. You can see general trends —summer is warmer than winter—and by studying today's weather patterns meteorologists can predict fairly accurately what is going to happen tomorrow, and perhaps even the next day. Beyond that, however, the system becomes steadily more unpredictable, because the tiniest variation in today's wind speed, temperature or air pressure can lead to vast changes in a few days. Bigger and faster computers improve things, but will never be able to solve the problem completely.

Challenge: Try your own chaotic patterns in the kitchen sink. Are big spoons better than little ones? What's the difference between the front and back of the spoon? Can you set up horizontal ripples in "standing waves"?

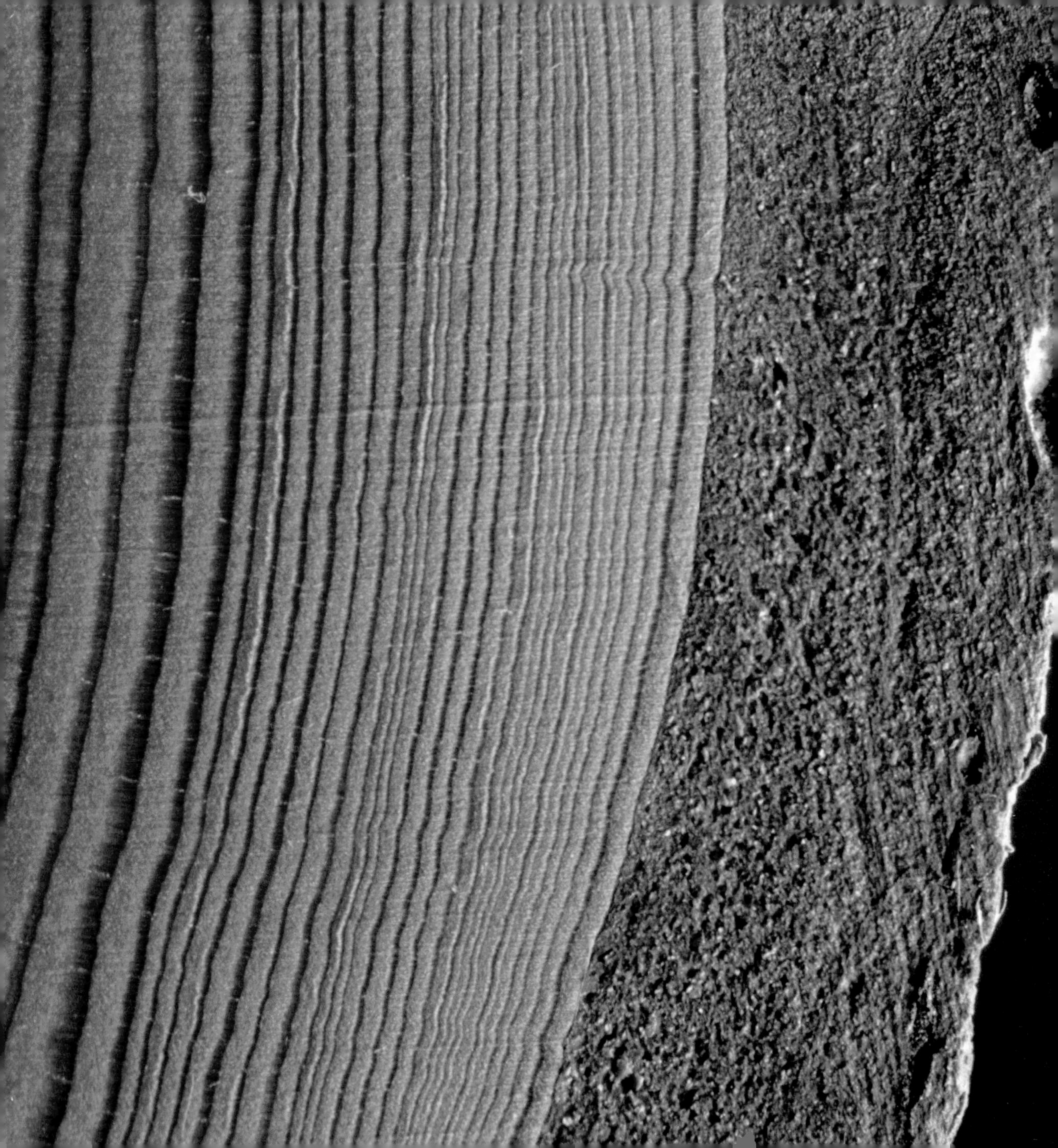

52. What is acid rain?

When water vapor condenses into clouds and forms raindrops, they are made of pure water. As they fall, however, they may pick up all sorts of things from the atmosphere, including dust, bacteria and soluble gases that are floating around.

There is a small amount of carbon dioxide gas in the air, and a little will be dissolved in the rain to make carbonic acid, but this will be so dilute as to be almost undetectable. Much more important are the oxides of sulfur and nitrogen.

Thousands of tons of coal are burned in power plants and other places. Coal is mostly carbon, but it usually contains a little sulfur. This burns to make sulfur dioxide gas, which dissolves in rainwater to make sulfurous acid. That may react with oxygen in the air to make sulfuric acid.

Nitrogen gas forms three-quarters of the atmosphere, and under normal conditions is unreactive. However, under abnormal conditions it can react with oxygen to make nitrogen oxides, which dissolve in water to make nitrous and nitric acids. These reactions can be caused by lightning, and also under the high temperature and pressure in the engines of cars and trucks.

In places where there is plenty of atmospheric pollution from power plants, factories and motor vehicles, the air will contain these oxides of sulfur and nitrogen, and the rain is acidic.

Acid rain has several unwelcome effects. For one thing it erodes limestone and marble statues. All over Europe there are statues that were sharp and clearly defined for centuries, but have in the last hundred years become disfigured and worn; faces have lost their features and fine detail has disappeared.

There has also been considerable damage to trees. This tree was felled in Germany's Black Forest in 1989. Inside the bark, the outer rings are all squashed close together, showing how growth had been stunted in the previous 20 years while motor vehicle traffic increased enormously.

DID YOU KNOW?

Sulfuric acid and nitric acid are two of the strongest acids in chemistry.

Challenge: Test your rain. Scrub a copper coin shiny with sink cleaner. Lay it on a windowsill or wall and cover half with tape. After heavy rain, look at the difference between the exposed and protected parts.

53. What shapes will tessellate?

When you tile a kitchen floor or the walls of a shower, you probably use square tiles, because they are cheap and fit together to cover the whole surface without leaving any awkward gaps. In practice you may have difficulty at the edges if the width is not exactly the same as a whole number of tiles, and you will have more trouble if your walls, like mine, do not meet at right angles. But in principle squares cover the whole surface: in other words, they tessellate.

Hexagons also tessellate; they fit together neatly without leaving gaps. They are less practical for a floor, since they cannot fit cleanly up to the edges of a square or rectangular room, but they will completely cover an infinite mathematical surface that has no edges. See the honeycomb on page 190 for an example of natural hexagon tessellation.

Challenge: What shapes tessellate on the surface of a soccer ball?

Pentagons will not fit together without leaving gaps, at least on a flat surface, but combine them with hexagons, wrap around, and you can have a ball.

DID YOU KNOW?

All quadrilaterals will tessellate; that is, you can take any shape with four straight edges, cut lots of copies, and they will fit together to cover the surface. This photograph shows how this works for a non-square shape, and it will work for any shape with four straight edges.

Challenge: Will any old triangles tessellate? Hint: Any quadrilateral can be divided into two triangles by cutting between opposite corners.

54. What shape is a satellite dish?

These dishes form a radio telescope at Cambridge in England. They pick up radio signals from objects far away in space and transmit them to the astronomy lab for analysis.

Each dish is designed so that when it is focused on a particular point in space, the radio waves from that point are reflected onto the receiver in the middle—at the focus of the dish. This means that the receiver is in effect many feet across, and therefore much more powerful than a simple aerial. The point of having several dishes spread out in a line is that they behave like one giant dish, and therefore have much greater sensitivity than one dish on its own.

If you were to cut the dish vertically down the middle then the shape of the cut edge would be a parabola, which is a mathematical shape that was studied by the Greeks 2,500 years ago, and given the name parabola by Apollonius of Perga (third century B.C.). Curiously, the same shape gives the trajectory of any missile thrown into the air and acted on only by gravity. Air resistance or drag modifies the shape, but even the table-tennis ball on page 134 has bounces that are close to parabolas. Old-fashioned electric heaters with glowing bars have roughly parabolic reflectors to throw all the heat out into the room.

DID YOU KNOW?

TV satellites are in orbits 22,000 miles (35,000 km) above the Earth, because at that distance they orbit the Earth in exactly 24 hours; so they appear to remain stationary. These geostationary orbits are now named Clarke orbits after Arthur C. Clarke, who first put forward the idea of such satellites in 1947.

Spin the mathematical parabola around its axis and you get a three-dimensional dish called a paraboloid, like these telescope dishes. Satellite dishes for television reception are the same paraboloid shape, since they are focused on satellites out in space; with impeccable logic the French call them paraboles.

The reflectors in car headlights and flashlights are a similar shape: the bulb is a small source of light and the point of the reflector is to throw out a strong, narrow beam of light. In practice the beams need to spread a little sideways, so the reflectors have a slightly different geometry.

55. How do crystals grow?

Most pure solid materials can form crystals under the right conditions. Wood will not form crystals because it is a mixture of many materials, rather than one pure substance. Water forms crystals—we call them frost and snow. Carbon forms crystals—diamonds. The photograph shows crystals of alum, otherwise known as ammonium aluminum sulfate.

Good crystals can often be made by dissolving a solid substance in liquid, and then allowing the liquid to cool or evaporate until the concentration of the substance gets too high; then it will begin to crystallize. The more slowly the liquid cools or evaporates, the bigger the crystals may grow.

Alum crystals are easy to grow. Alum used to be a vital ingredient in dyeing cloth. Dip some wool in a solution of natural dye and it comes out colored, but the dye will wash out easily, because it does not bind to the wool. Alum is a mordant, meaning a chemical with teeth. It bites into the fibers of the wool, and it bites into the molecules of dye and holds them together. So soak your wool in a solution of alum and then dip it in the dye and the color will be brighter and fast—it will not wash out.

Challenge: Buy some alum from your drugstore. Dissolve a few spoonfuls in a small amount of hot water in a pan or bowl, stirring to make sure it all dissolves. Cover it to keep dust out, and leave it in a warm place for an hour or two. Crystals should start to grow. Then transfer the whole container to the fridge, and soon more crystals should grow. Pour away the remaining water and dry the crystals with paper towel. Don't wash them or else the crystals will begin to dissolve and lose their sharp edges.

DID YOU KNOW?

In Britain there is no alum lying around, but experimenters in the early 1600s found out how to make it. They dug gray shale out of the cliffs of North Yorkshire, roasted it slowly for nine months and washed the residue with water. They added stale human urine to the washings, heated the mixture until a fresh chicken's egg just floated to the surface, and then allowed it to cool. Beautiful crystals of pure alum, just like these, grew out of the solution.

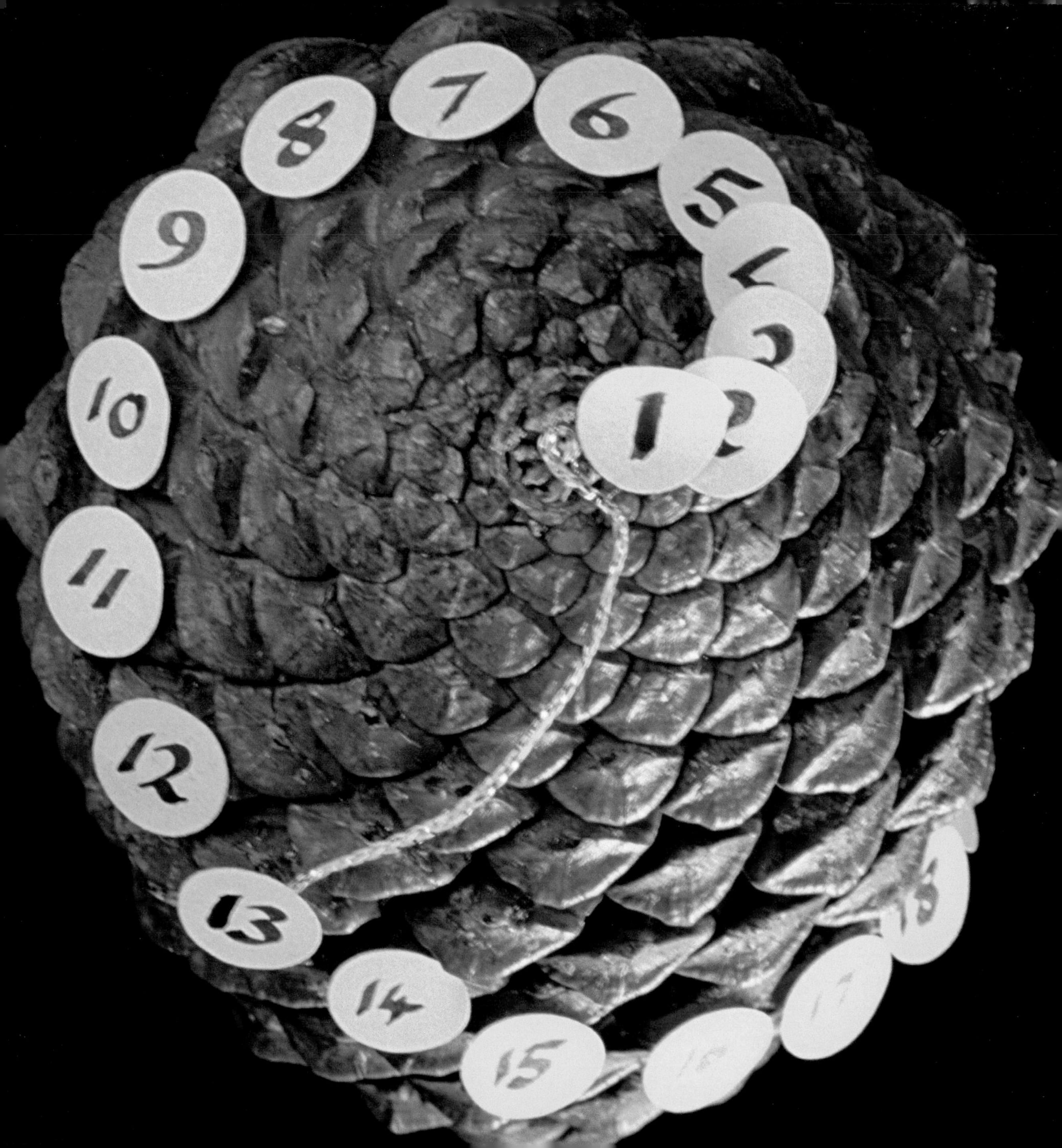
1
5
6
7
8
9
10
11
12
13
14
15

56. Who was Fibonacci?

Leonardo of Pisa (c. 1170–c. 1250), who lived in Pisa while the first few layers of the leaning tower were being built, was generally known as Filius (son of) Bonacci, or Fibonacci for short. He was a gifted mathematician, and he traveled around the Mediterranean learning from the Islamic scholars who were then the best mathematicians in the world.

In 1202 he wrote a book, *Liber Abaci*, in which he introduced "Arabic numerals" to the West. He wrote: "These are the nine figures of the Indians: 9 8 7 6 5 4 3 2 1. With these nine figures, and with this sign 0 ... any number can be written, as will be demonstrated."

In the same book he posed a problem about rabbits. Suppose you start with a pair of baby rabbits; they take a month to mature and after that have a pair of baby rabbits every month. Assuming that all the rabbits are equally productive, how many pairs will there be at the end of each month? The answer turns out to be the sequence 1, 1, 2, 3, 5, 8, 13, 21...

Challenge: Can you work out the next three numbers in the sequence?

This Fibonacci sequence is curious in several ways. Add the first and the third numbers together. Try the first and the third and the fifth, or the first and the third and the fifth and the seventh; what do you get?

The ratio between any two consecutive numbers—say 5/8 or 8/13, gets closer and closer to the golden ratio 0.618 the further along the sequence you go. The golden ratio itself is something of a mathematical curiosity, because $0.618 = 1 \div 1.618$, and it often seems to turn up in paintings and architecture.

Meanwhile, the numbers keep cropping up in nature. Many flowers have not four petals, as you might expect, but 3 or 5 or 8 or 13; the daisy on page 159 has 21 petals. The sea urchin shells on page 182 have five-fold symmetry.

The leaves of thistles grow in spirals; so do sunflower seeds and many plants with branches off the stem. Often the spirals repeat every 8 or 13 up the stem. This pine cone has 13 spirals running clockwise and 8 counterclockwise. From the top down to where the marked spirals cross again is 8 lumps clockwise and 13 counterclockwise.

Can this all be coincidence?

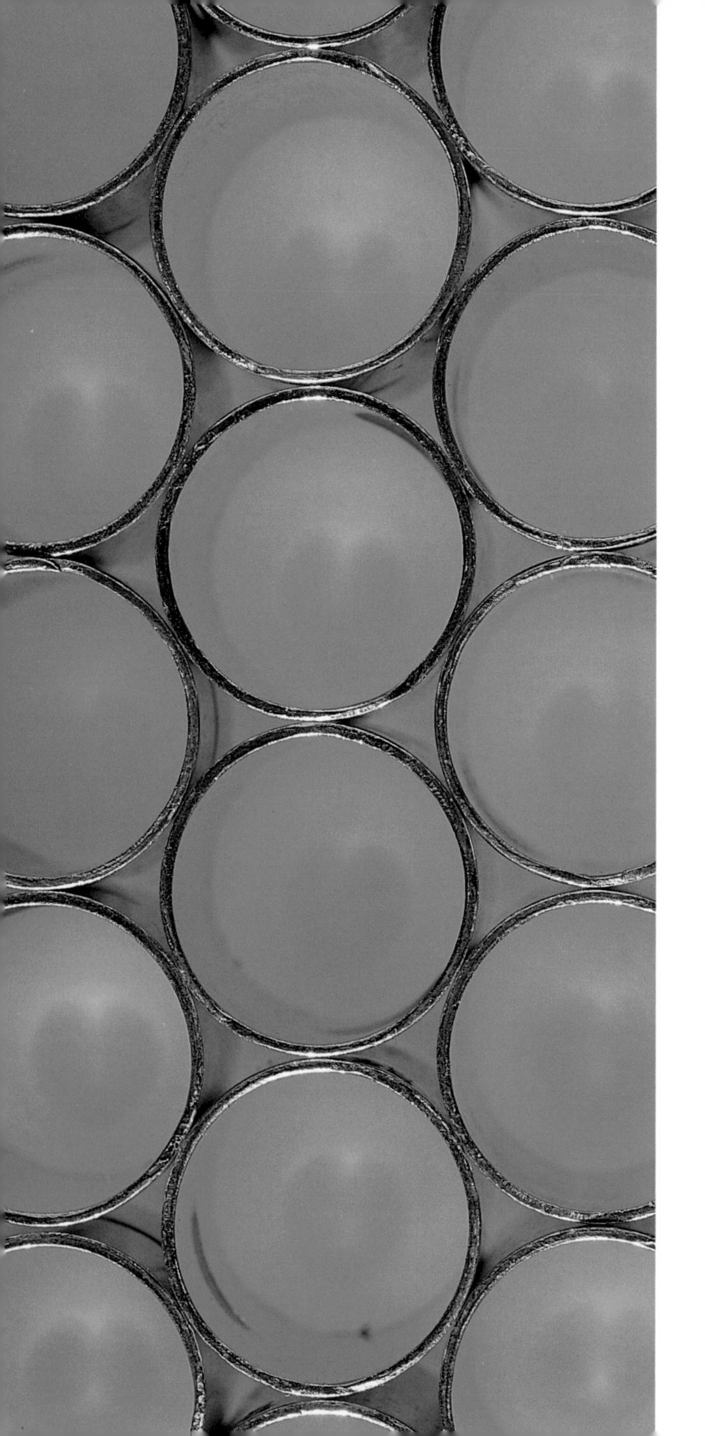

57. Which numbers are angular?

The ancient Greeks seem to have been the first people to study numbers, to love them, and to actively look for number patterns. They had almost certainly investigated triangular numbers by the time of Pythagoras, around 550 B.C., perhaps not with numerals but with pebbles in the sand.

Pythagoras himself was obsessed by numbers. Born on the island of Samos, he eventually set up a curious mathematical school at Croton, in the "instep" of what is now Italy. They were a secretive group, and had a range of strict rules; they were not allowed to touch white feathers, eat beans or "make water in the sunshine."

For them, numbers were everything, and they assigned particular numbers to stand for what was important—1 for the beginning, 2 for Earth, 8 for love, and so on. Pythagoras was also interested in music. He probably laid the mathematical foundations of the musical intervals that we use today, and he invented the "harmony of the spheres," which is the notion that the gaps between the five planets then known were proportional to the gaps between the principal notes of the musical scale.

There is a neat connection between triangular and square numbers: add any two consecutive triangular numbers and you get a square; so 3 + 6 = 9 and 6 + 10 = 16. Conversely, every square number is the sum of two successive triangular numbers.

Hexagonal numbers are useful in particular stores: people often sell cardboard tubes in bundles of 7 or 19 or 37.

Challenge: Can you explain why?

The angular numbers were simple numerical patterns. A triangle with one row has one pebble, with two rows three pebbles, with three rows six, and so on. So the sequence of triangular numbers goes 1, 3, 6, 10 …

Challenge: Can you continue the sequence?
The answer: simply add successive whole numbers; so the fifth triangular number is 1 + 2 + 3 + 4 + 5.

The square numbers are 1 (1 x 1), 4 (2 x 2), 9 (3 x 3), etc. To get them, as you can see in the knitting, you add odd numbers: 1 yellow; 1 yellow + 3 green = 4; 1 yellow + 3 green + 5 black = 9, and so on.

Challenge: Can you continue the sequence?

12kg
500 g ℮

58. How big is a million?

Once, for a television program I was producing, I phoned a sugar company and asked for a million sugar cubes. They generously sent us 10,000—here are some of them.

Challenge: Suppose the company had sent us a million sugar cubes, how could they have delivered them? Would they have needed a bike? A van? A truck?

Each sugar cube was 1 cubic centimeter ($^{1}/_{16}$ cu. in.)—1 cm x 1 cm x 1 cm. A million would stack into a block 100 cubes long, 100 cubes wide and 100 cubes high, because 100 x 100 x 100 = 1,000,000. But 100 centimeters is 1 meter; so the entire stack of a million cubes would fit exactly into a cubic meter—it would fit under a desk.

The density of cubed sugar is about 1.5 g/cm^3 or 1,500 kg/m^3; so a cubic meter of sugar cubes weighs about 1,500 kg, or 1.5 metric tons. You could not carry it on a bike, but a small van would do. In practice it would be extremely tedious to stack the little cubes into a big cube without any air spaces, and the cubes they sent were all jumbled in the boxes with lots of air gaps. A million would still have weighed only 1.5 tons, but would have occupied most of a 5-ton van.

Challenge: Suppose you laid the million cubes out on the ground, all touching, to make a solid square sheet of sugar, one centimeter thick. How much space would it occupy? Would you need a large room? A tennis court? A football field?

A million is a thousand thousands: 1,000,000 = 1,000 x 1,000. Therefore the flat square would have a thousand cubes along each side, and a thousand cubes make 1,000 x 1 cm = 10 m. So the sheet would be just 10 meters square; you would need a large room or a tennis court.

Challenge: What if you piled up the million sugar cubes one on top of another? How high would the pile be?

All the cubes are in one vertical line. The height of the pile is therefore a million centimeters: 1,000,000 cm = 10,000 m = 10 km. So the pile would be 10 kilometers high, slightly higher than Mount Everest (8,850 meters).

59. How long is a centimeter?

For thousands of years people have wanted to measure things. For most of that time they used parts of the body to measure with: the length to the first joint of thumb or forefinger was an inch, the length of the foot was a foot, and for the Romans a mile was a thousand *(mille)* paces.

Horses are still measured in hands (1 hand = 4 inches/10.16 cm), and some other parts of this system—inches and miles, for instance—are still used today, but it does have drawbacks. One simple difficulty used to be the lack of standards. In about 240 B.C. a Greek mathematician named Eratosthenes worked out the circumference of Earth, and we think he was pretty accurate, but we don't know how accurate because he measured it in stades, and we aren't sure how long his stade was.

This problem became acute in France in the late 1700s, when a great variety of measures were being used in markets, and some of them varied by 30 percent from one village to the next, which made life almost impossible for merchants and traders. Cloth was measured in aunes, but there were three different aunes in Paris for different types of cloth, and at least a dozen others in just one small group of villages.

After the French Revolution, Napoleon (1769–1821) decided that all his people should use the same weights and measures, and he commanded French scientists to set new standards for the world. He invented a new measure of length, the meter, which would be $^1/_{10,000,000}$ of the distance from the North Pole to the Equator, going through Paris. The surveyors spent eight years making the measurements, and even then they got the answer slightly wrong, but nevertheless we now have a standard meter. A centimeter is $^1/_{100}$ of a meter—about the width of a little fingernail—and a millimeter $^1/_{1,000}$ of a meter (as thick as a couple of fingernails).

DID YOU KNOW?

The standard meter used to be the length of a metal bar in Paris, but that is now obsolete. Since 1983 the meter has been defined as the length of the path traveled by light in a vacuum during a time interval of $^1/_{299,792,458}$ of a second.

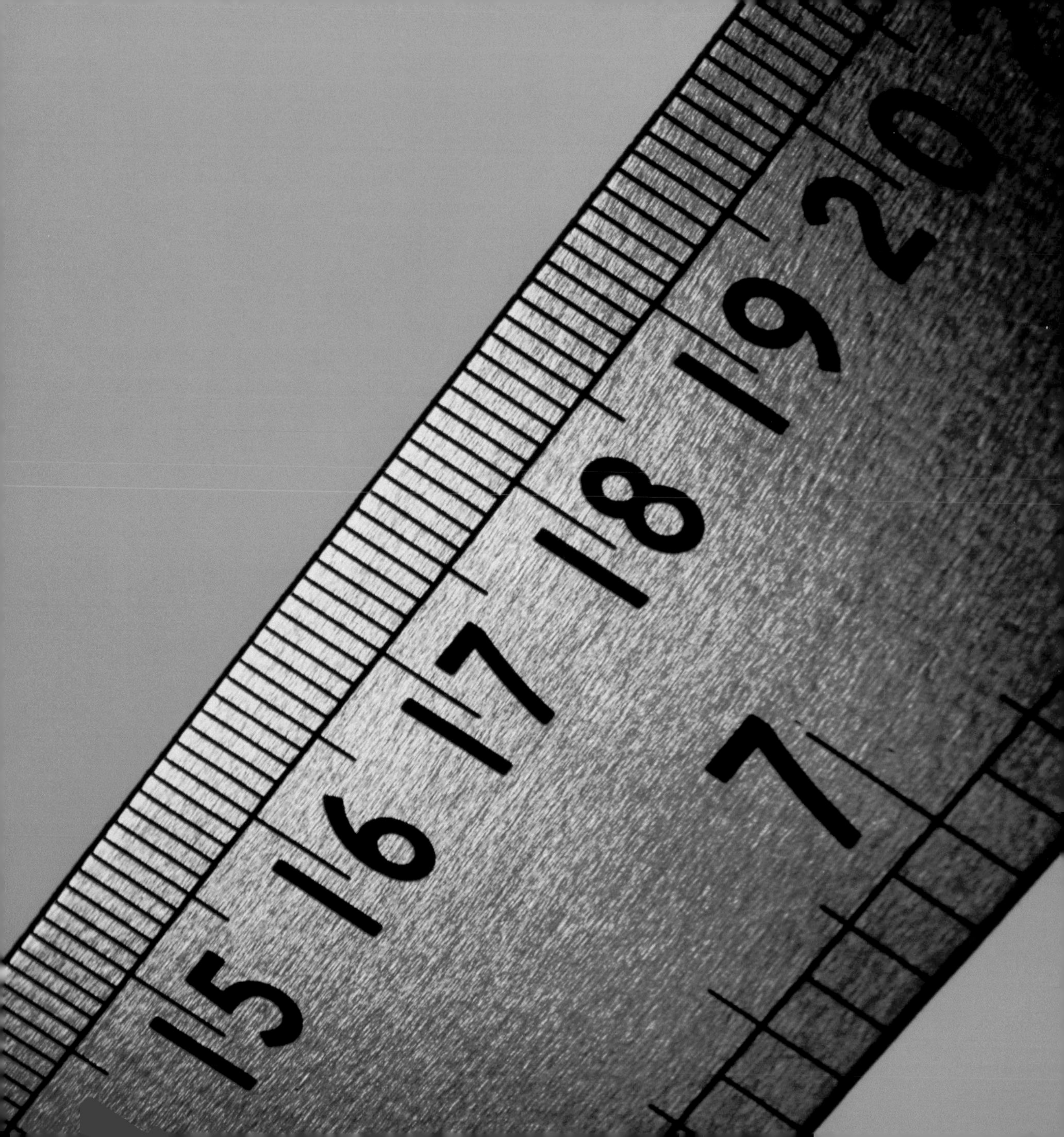

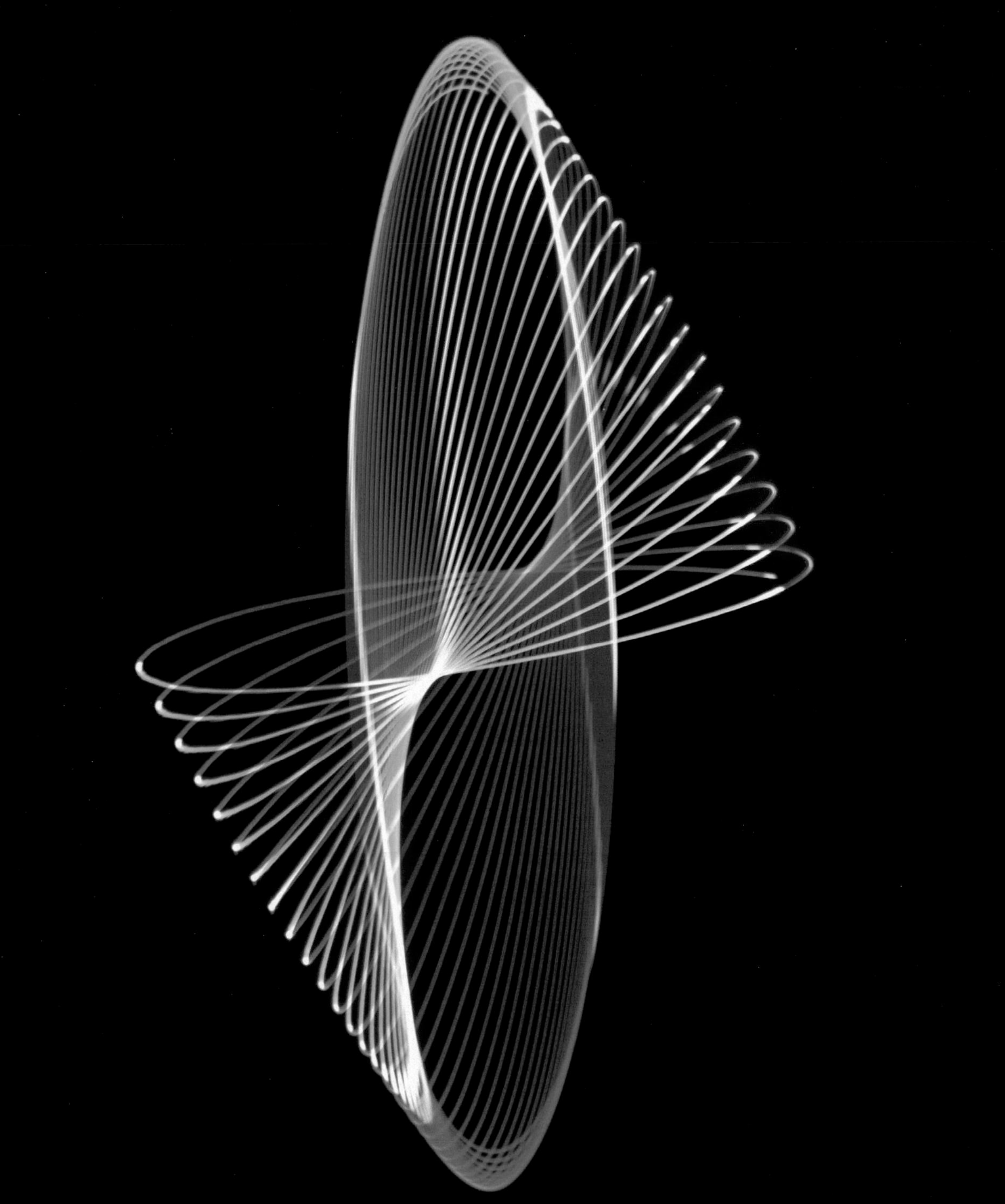

60. How does a pendulum swing?

According to legend, the basic features of a pendulum were first recorded by Galileo Galilei (1564–1642), in the cathedral at Pisa, where he was a medical student. Bored by a tedious sermon, he watched the swinging of the huge bronze lamp suspended on a long chain. Sometimes it swayed slightly, and sometimes it swung in a wide arc, driven by a drought. Using the pulse in his wrist, he timed the swing of the lamp and discovered that regardless of the width, it always took the same time for one complete swing.

DID YOU KNOW?

The time of one complete swing—the period—depends neither on the width of the swing, nor on the weight of the bob, but only on the length of the pendulum. The period is proportional to the square root of its length in metric measurements; so a pendulum 1 m (3¼ ft) long swings twice as fast as one 4 m (13 ft) long, and three times as fast as one 9 m (29½ ft) long.

Galileo devised a neat little gadget with a pendulum whose length could be varied from about 4 to 8 inches (10 to 20 cm), calibrated it, and then used it to measure patients' pulse rates. In those days—long before accurate watches existed—this was a brilliant invention. The medical school rapidly took up the idea and developed it, but poor Galileo received no recognition.

Galileo realized that a pendulum could form the timekeeper for an accurate clock; he designed one, but did not get around to building it. The first person to make a pendulum clock was the Dutch scientist Christian Huygens (1629–95), and with improvements by Robert Hooke (1635–1703) the pendulum clock was the most accurate type of timepiece for 300 years. Robert Hooke also invented the universal joint, which he used to make what he called a "conical pendulum." This could swing both north–south and east–west, but could not twist. He used this to investigate the orbits of planets and moons, since the bob can swing in a circle or an ellipse. The photograph here is a harmonogram—the complicated result of the combined motion of two conical pendulums, and so represents a tribute to Galileo, Huygens and Hooke.

Challenge: Make a two-pendulum harmonograph and see what patterns you can get with pen on paper.

61. How fast does a pendulum swing?

Here are 26 images of a brass weight hanging on a piece of string and swinging from left to right—in other words, it's a simple pendulum in action. The driving force is gravity: the weight starts high on the left, and gravity pulls it downward. It can't fall straight down, but it can get to a lower place by swinging.

As it falls it accelerates, until it gets to the lowest point. After this the weight is moving upward, so gravity slows it down and it comes to rest opposite the point at which it started. Captured every 50th of a second, the images of the weight and string are close together in the green sections at the beginning and end of the swing, where movement is slow, and further apart in the middle, orange section, where movement is fast.

The speed of the swing follows the rule of 12, which is common in natural rhythms. Divide the total angle of swing into six equal sectors, and the speed in each sector is given roughly by dividing 12 unequally into six:

Sector	A	B	C	D	E	F
Speed	1	2	3	3	2	1

The same rule of 12 applies to the tides, which come in and go out roughly every 12 hours. At high tide and at low tide the water moves slowly up the rocks and across the sand, but at half tide the water moves rapidly—in some places so rapidly that people have been overwhelmed and drowned. So if you want to swim across an estuary, the safest time to start is at "slack water," which is just before high tide or, even better, just before low tide, when the channel may be narrower. That's when there is the least current flowing across your path.

If you get up early and watch the first sparkle of the sun appear, you'll soon see that from day to day it moves along the horizon. Watch for three months, and you can see the same rule of 12. In midwinter and midsummer, the sun stands still—that is what "solstice" means—and turns slowly back, while at the equinoxes in March and September the sun seems to leap along the horizon; the first flash changes considerably from day to day.

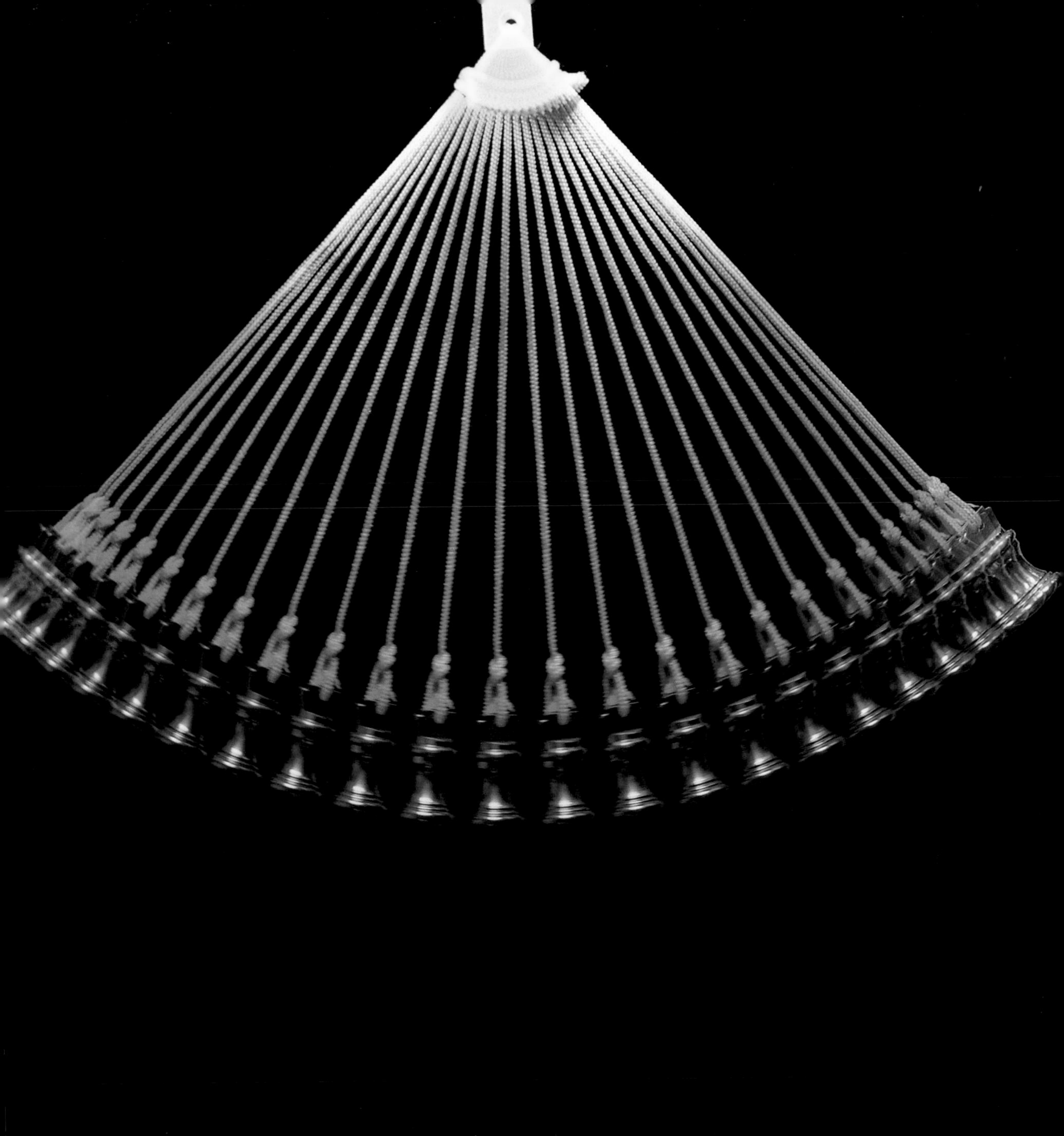

technology

62. Why don't knots come undone?

This knot is a double sheet bend, recommended for joining a thin rope to a slightly thicker one. There are two main kinds of knots: bends, like this one, are for joining two ropes, while hitches are for tying, or hitching, a rope to a post or other solid object.

Knots have always been used on boats, for example, to tie the boat to a mooring, a jetty or an anchor, to raise, control and reef the sails, and so on. In domestic onshore life, knots are still used extensively, from neckties to shoelaces.

Knots have evolved over hundreds of years as people have investigated every possible way to tie things, and the ones that survive, on board ship and in the knot books are the "fittest" in Darwinian terms—the ones that are most useful. The best knots are easy to tie and untie, but do not slip when you don't want them to.

DID YOU KNOW?

The only thing that stops knots from coming undone is friction.

Twisting rope or string around itself pulls the fibers together, and so the friction is high. Furthermore, good knots are self-tightening: tension tightens the knot, increasing the frictional force that prevents it from slipping.

All knitting and weaving depends on the same idea. Without friction a beautifully knitted sweater would unravel at the slightest snag. The warp and weft of woven fabrics would simply slide off one another, and the most elegant suit would rapidly develop gaping holes. Clothes are held together by friction.

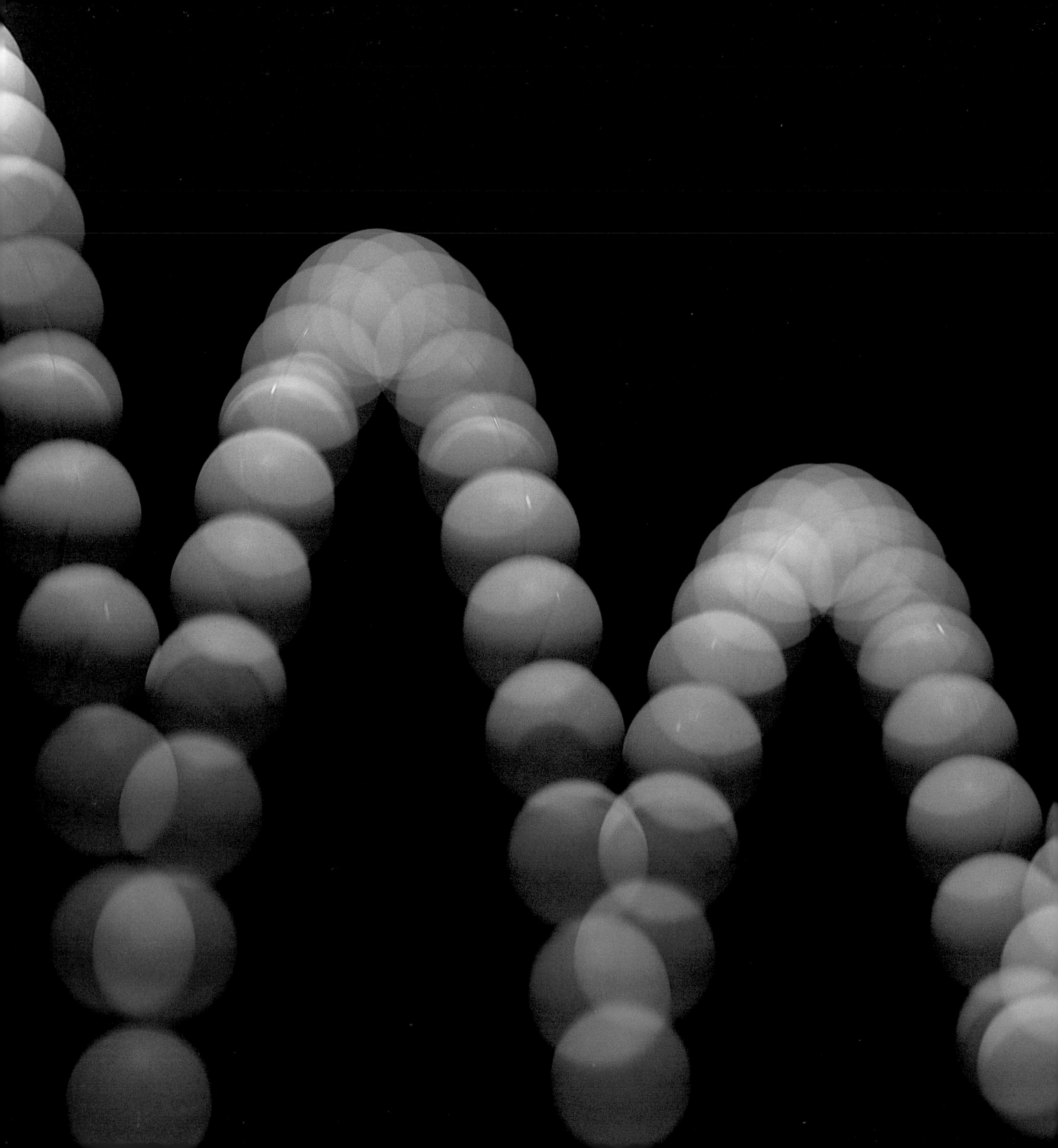

63. Why does a ball bounce?

As a ball falls, it accelerates and picks up kinetic energy. When it hits the ground, most of that kinetic energy is stored in compression of the ball and the surface; in other words, converted to potential energy. In the case of a table-tennis ball, the celluloid is slightly flattened, and the gas inside is compressed. When all the kinetic energy has been transferred, the ball stops moving downward, and the compression of both ball and surface begin to lift the ball again. The surfaces have been deformed, even if only by a minute amount, and they begin to spring back. The resulting force kicks the ball back into the air. Each time the ball hits the surface it bounces off at almost the same angle on the other side of the vertical.

This entire multiple bounce took just over one second. At the beginning, top left, the images are almost on top of one another, because the ball is moving slowly. As it falls down the left side of the page you can see the images getting steadily further apart as the ball accelerates. If there were no air resistance and no surface in the way, the ball would, after one second, reach a speed of nearly 32 feet (10 m) per second.

DID YOU KNOW?

No ball ever bounces as high as the place it fell from. If it bounced higher you could extract energy from it and make a perpetual motion machine. Unfortunately, the second law of thermodynamics says that some energy will always be lost, in this case to sound and heat. For the same reason the second bounce is lower than the first; indeed, each bounce is lower than the one before.

After the first bounce, the images are far apart although not as far apart as just before the bounce, because of the lost energy. Gradually, they get closer and closer together, until at the top of the first bounce they are almost on top of one another and the ball is moving neither up nor down, but only from left to right.

The shape of each bounce is roughly a parabola, which if there were no air resistance would describe the trajectory of any unpowered missile—a baseball, a soccer ball, a football or an artillery shell.

64. What are machines?

The word "machine" has been used to describe any structure, implement or vehicle, from snowshoes to ships, but now usually means "an apparatus for applying mechanical power, consisting of a number of inter-related parts, each having a definite function" (*Oxford English Dictionary*). Many machines have gears, whose function is to change the rate of rotation. Drive a cog-wheel that has 40 cogs or teeth with one that has only 20, and the larger cog-wheel will rotate at only half the speed of the smaller one.

The earliest gears were made of wood; they were built long before metals were readily available. In mills driven by wind or water-wheels, power from the main shafts has to be transferred, to lift the sacks of grain, to turn the grindstones, and so on. Crude wooden gears also converted the slow circulation of donkeys or oxen into useful motion elsewhere. These machines, mainly of wood, were the province of the millwright.

Next came the clockmakers, who started with wood but progressed to metal, and had to make spindles and gears with precision so that their clocks would keep time accurately. The first good mechanical clocks appeared in the 13th century, and their quality went on improving until the late 20th century, when timekeeping was taken over by electronics.

Meanwhile, in early 18th-century Britain, two great inventions revolutionized machines. Abraham Darby (c. 1678–1717) worked out how to make cheap iron, using coke rather than charcoal, and Thomas Newcomen (1663–1729) built the first useful steam engine for pumping water out of coal mines. The Industrial Revolution was slow to take off, but by 1776 engineer Matthew Bolton (1728–1809) was boasting to Scottish man-of-letters James Boswell: "I sell here, Sir, what all the world desires to have—Power." Bolton had vast iron machines and a workforce of 700 men to run them, but in fact his boast was scarcely true at the time, since James Watt (1736–1819) had spent 10 years getting his first steam engine to work, and in 1776 the business had not begun to make money. In the long run he was right, though, for by the time they retired at the end of the century, Bolton and Watt had become rich—from the application of mechanical power—from their machines.

3 2
2 1 0
2 1 0
3 2
0 1
1 2
5 6
9 0
5 4
8 7

65. What was the first precision device?

While still a student, Charles Babbage (1791–1871), a difficult and irascible man, became obsessed with the inaccuracies in mathematical tables, and worked out a way of calculating them mechanically. He designed what he called a "Difference Engine," and hired a skilled engineer, Joseph Clement (1779–1844), to build it. This photograph shows the remains of the demonstration model of Babbage's Difference Engine, built in the 1820s.

Unfortunately, the full-size Difference Engine was never completed until Science Museum engineers got to work 170 years later and built it to Babbage's specifications. I have seen this wonderful beast in action twice, and both times it jammed. Babbage's machine probably would never have worked, even if it had been finished, since the precision required for all those brass components was beyond what was possible at the time.

Babbage is often hailed as the father of this sort of technology, but he was not the first. Samuel Morland (1625–95), a spy who, after saving the life of Charles II, was knighted and made Master of Mechanicks, built an adding machine in the 1660s, but it was not automatic; diarist Samuel Pepys dismissed it as "really rather silly." Morland had probably seen the work of Blaise Pascal (1623–62), the brilliant French mathematician who invented probability theory to solve a betting problem. In the early 1640s, Pascal made a calculating machine that could add and subtract automatically.

However, pride of place must go to the unknown Greek craftsman who, around 100 B.C., built a machine with 32 bronze cog-wheels in a wooden box, designed to predict the movements of the planets. This amazing device was being taken from Kos toward Rome when the ship carrying it sank off the island of Antikythera. Known as the Antikythera mechanism, this is the oldest known precision device in the world. It is an astonishing demonstration that the ancient Greeks were as skillful with their tools as they were with their philosophical minds.

DID YOU KNOW?

Babbage hated street music, and succeeded in getting a law passed to ban busking. This was very unpopular; a dead cat was thrown at him in the street and bad musicians were paid to play loudly outside his house.

66. Which are the world's oldest toilets?

In 1850 a severe storm ripped the turf off sand dunes at Skara Brae, on the west coast of the mainland of Orkney, just north of Scotland. Underneath were revealed eight Stone Age houses, huddled down out of the weather, and built some 5,000 years ago.

These houses form a Neolithic housing complex. They are close together—separated only by narrow passages—and they all seem to be almost identical in design. Each house has just one room, about 270 square feet (25 sq. m). In the center (brown in the photograph) is an open fireplace, used for heating and cooking. The door is in the middle of one wall—underneath the near left-hand side of this picture—and opposite is a large stone dresser, presumably there to display valuable possessions. Luckily for us, there seems to have been a shortage of wood, and so the furniture was built in stone and has therefore survived.

Between the "dresser" and the fireplace is a seat for a distinguished visitor—far from the drafty door. On the right is a large box bed, and on the left a small one. These may have been for dad and mom respectively, or possibly the large one was for dad and mom, and the small one for the children. Above each bed are cupboards in the walls—bedside shelves, although they had no books, glasses, remote controls or bedside lights to put there.

Near the dresser are three small boxes, sunk slightly into the ground, and waterproofed with clay. The best guess is that these were bait boxes. These people must have eaten a great deal of fish, for there are hundreds of limpet shells in their trash piles. Limpets do not make good eating—I am reliably told that they are tough and taste like slightly fishy rubber—but they do make good bait for fishing, especially if they have been soaked for a day or two. So perhaps these boxes were full of soaking limpets: fresh, one day old and two days old.

Best of all, in the corner of each house is a small doorway leading into an alcove perhaps 2 feet square (70 cm sq.), with a hole in the middle of the floor. The hole connects to network of drains that go underneath all the houses and down toward the sea. These neat ensuite cubicles are the oldest toilets in the world.

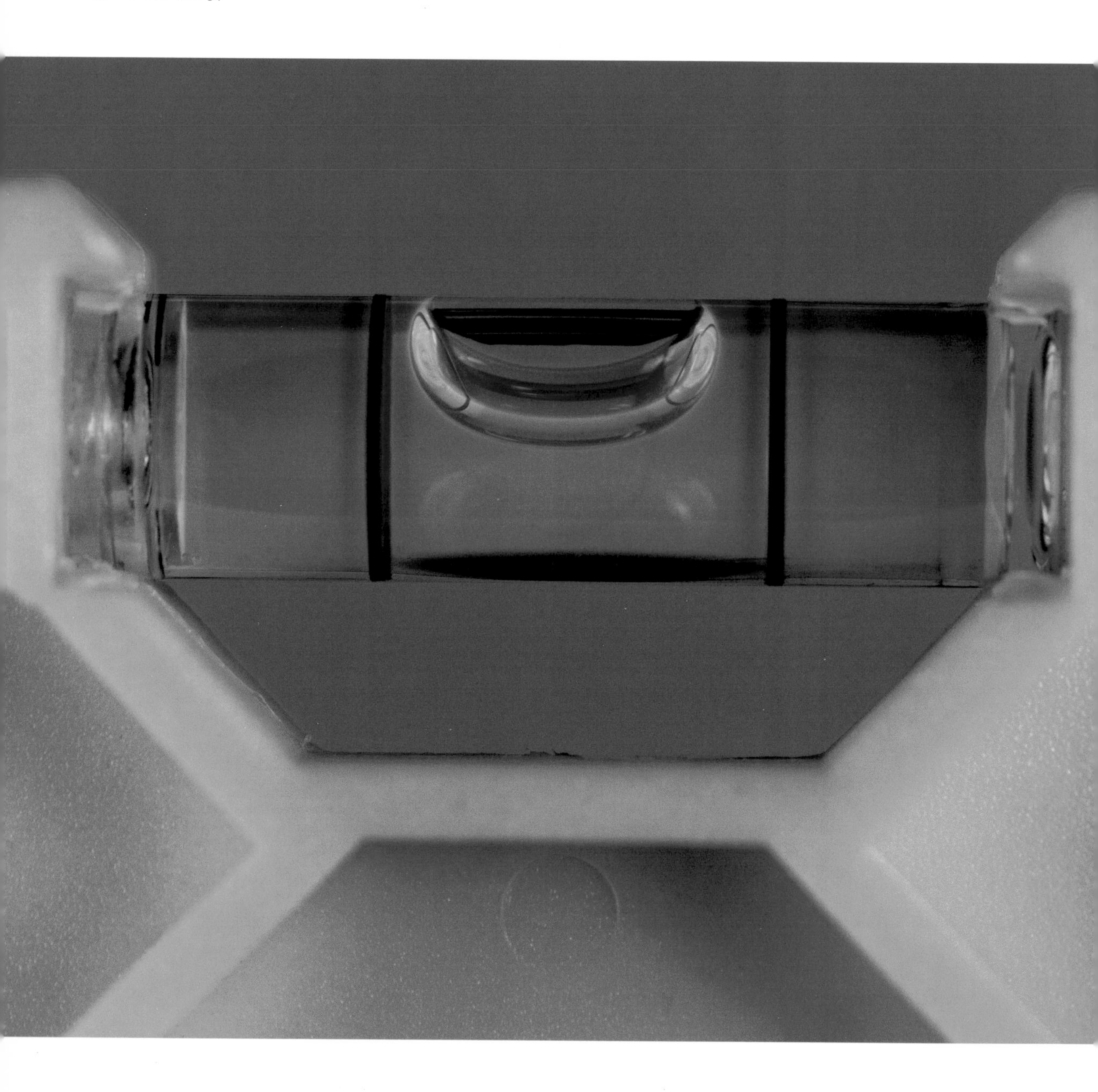

67. How do you get a level playing field?

Challenge: What do you understand by "level"?

There are two ways of thinking about it. First, gravity pulls things straight down toward the center of the Earth, so you can find a true vertical by using a plumb bob (a weight hung on a piece of string). Any line at right angles to the string is level, or horizontal. Just watch out for the attraction of mountains (see page 40).

Builders need to get their structures horizontal and vertical; otherwise things tend to go wrong—buildings tend to look lopsided and may possibly fall over. Builders use plumb bobs to find a true vertical and levels to find a true horizontal. As John Smeaton (1724–94) wrote of his Eddystone lighthouse in 1793: "It could be brought justly horizontal by means of a pocket Spirit-Level being placed upon it."

The level has a flat base; above it is a glass tube filled with alcohol plus a bubble of air. The glass tube is very slightly curved, so that the ends are the same distance from the base but the center is fractionally higher. Lay the level on a horizontal surface and the bubble will come to rest exactly in the middle of the tube, but if the surface is not level the bubble will be off center.

DID YOU KNOW?

The word "horizontal" provides another clue; it means parallel to the horizon. The horizon is really curved, because Earth's surface is curved, but if you stand by the sea and look over the water the horizon is very nearly flat, and its line is horizontal.

Challenge: Can you explain why the bubble comes to the center when the level is on a horizontal surface?

When builders need to establish a level between two places with an obstacle between—say around the corner of a garage—they can use the fact that water always finds its own level. Lay a section of garden hose from one place to the other, propping both ends up to the level you wish to check. Pour water carefully into one end until the hose is full. If the ends are level the water will reach both, but if not, then the water level will be below one end of the hose.

68. Why are sand ripples interesting?

Phoebe Sarah Marks (1854–1923) was a tough girl. The daughter of an immigrant Polish Jew, she disliked rules, taking no for an answer and her name, which she changed to Hertha. She was clever and went to Cambridge University, and then to Finsbury Technical College to work with Professor William Ayrton (1847–1908), whom she married the following year, 1885.

She joined in her husband's research into electric arcs, but the paper they wrote was destroyed by a servant, who used it to light the fire. William moved on to other things, but she started again, improving the technique. She became the world expert on the subject, and the first woman member of the Institute of Electrical Engineers. But because she was a woman she was not allowed to read her paper to the Royal Society; it had to be read for her by a colleague.

In 1901 she went on holiday to Margate in Kent, and there on the beach she became intrigued by the ripples in the sand, which seemed to form at right angles to the waves. What could be going on? Perhaps a slight irregularity—a pebble or seashell—set up a vortex that produced a hollow, and a little heap of sand an inch or so further on—and the same thing on the way back.

She worked out the mathematics, wrote an elegant paper, and overthrew a male bastion that had stood for 240 years, for in 1904 Hertha Ayrton became the first woman ever to read a paper to the Royal Society. She also became a suffragette and campaigned to get votes for women.

During the First World War, the soldiers in their trenches came under the terrible threat of poison gas, and the race was on to develop antidotes, effective gas masks and ways to get rid of it. Hertha Ayrton took her sand-ripple mathematics and ran it backward; a vortex at one end should create a wave at the other. Turning the idea into practice, she designed a fan, made from wood and canvas, that could be used to drive the poison gas out of the trenches.

In all, 105,000 Ayrton fans were made and dispatched to the troops in France, although sadly many, arriving without instructions, provoked only bewilderment and were used as firewood.

MUNI QUIP K-GP

69. How do radar guns work?

When a police car or ambulance comes zooming toward you on the road, the note of the siren seems to change from high to low as it passes you. This sudden drop is called the Doppler effect, after Christian Johann Doppler (1803–53), who described it 1842.

The source—siren or trumpet—puts out only one pitch; the note does not really change, but you hear it change because of the speed of the vehicle. The sound waves are traveling from the source to your ears, and when the source is moving toward you the waves are in effect being compressed. More crests reach you every second, which is equivalent to a higher note.

When the source is moving away, the sound waves are stretched; fewer crests reach your ears each second, and you hear a lower note.

Radar guns use the same principle. The gun sends out a beam of radar waves at a particular frequency (or "note"). The waves bounce back from a car, and the echo is picked up by the gun and compared with the original. If the car is stationary, the frequency of the echo is the same as that of the outgoing signal, but if the car is moving, then the frequencies are different. From the difference, the gun can calculate the speed of the car.

Medical scientists use a similar technique to study blood flow, although with ultrasound rather than radar. When there is a problem with the blood supply to your brain, you may get a carotid artery scan. A small ultrasound gun held against the carotid artery in the side of your neck can measure the flow rate of blood inside, without having to make a mark on your skin.

The Doppler effect also works with light waves—the light is shifted toward higher (blue) or lower (red) frequencies—and has been used to measure the sun's speed of rotation. This has revolutionized cosmology; in 1929, Edwin Hubble (1889–1953) used the redshift of stars to show that the universe is expanding.

DID YOU KNOW?

In a spectacular demonstration, Doppler arranged for a group of trumpeters to play in an open railway carriage while speeding past a group of musicians with perfect pitch, and they endorsed his observations.

70. How hot are your lips?

This image is a thermogram, or heat image, and was taken with a thermocamera, or thermal imaging camera. Instead of visible light, this camera responds to heat, or infrared radiation.

The colors represent temperatures; the highest are white and yellow, the lowest purple and black. So the hottest parts of the couple are his cheek and her neck, probably because of the blood flow close to the skin. The noses are cooler, because they are more exposed to the air, which takes the heat away—that is why your nose gets cold when you go out on a cold day. Their lips don't appear to be noticeably hot.

Clothes clearly provide good insulation, keeping the heat in, and therefore look cool in the picture. Surprisingly, hair seems to provide almost as much protection.

Engineers use similar equipment to look for hotspots in industrial machinery. When some part of a machine is worn or out of alignment, there may be excessive friction that could eventually lead to trouble. The thermal image will show up any hotspots before the trouble starts, and the engineers can take appropriate action. Thermal images are useful for non-destructive testing of chains under load or other systems under stress, and for checking boilers and reactors for damage to the thermal insulation. A thermogram of a building will show where it is losing heat, and whether the roof or the windows need better insulation. For the same reason, security services use thermocameras to check whether there is anyone hiding in an apparently empty building.

At a more intimate level, close-up thermal images can reveal hotspots in body tissues caused by inflammation. Such hotspots are useful in trying to pinpoint the causes of lower back pain and muscle strain, for example. Thermography is helpful for doctors looking for breast cancer; two breasts showing different temperatures or the pattern changing over time may indicate the presence of a tumor. For veterinarians thermograms can be invaluable, since animals cannot tell them where it hurts.

DID YOU KNOW?

After a building has collapsed, maybe due to an earthquake, emergency services use thermal imaging cameras to look for people trapped in rubble, because even unconscious people emit heat, and this can show up through layers of dust and concrete.

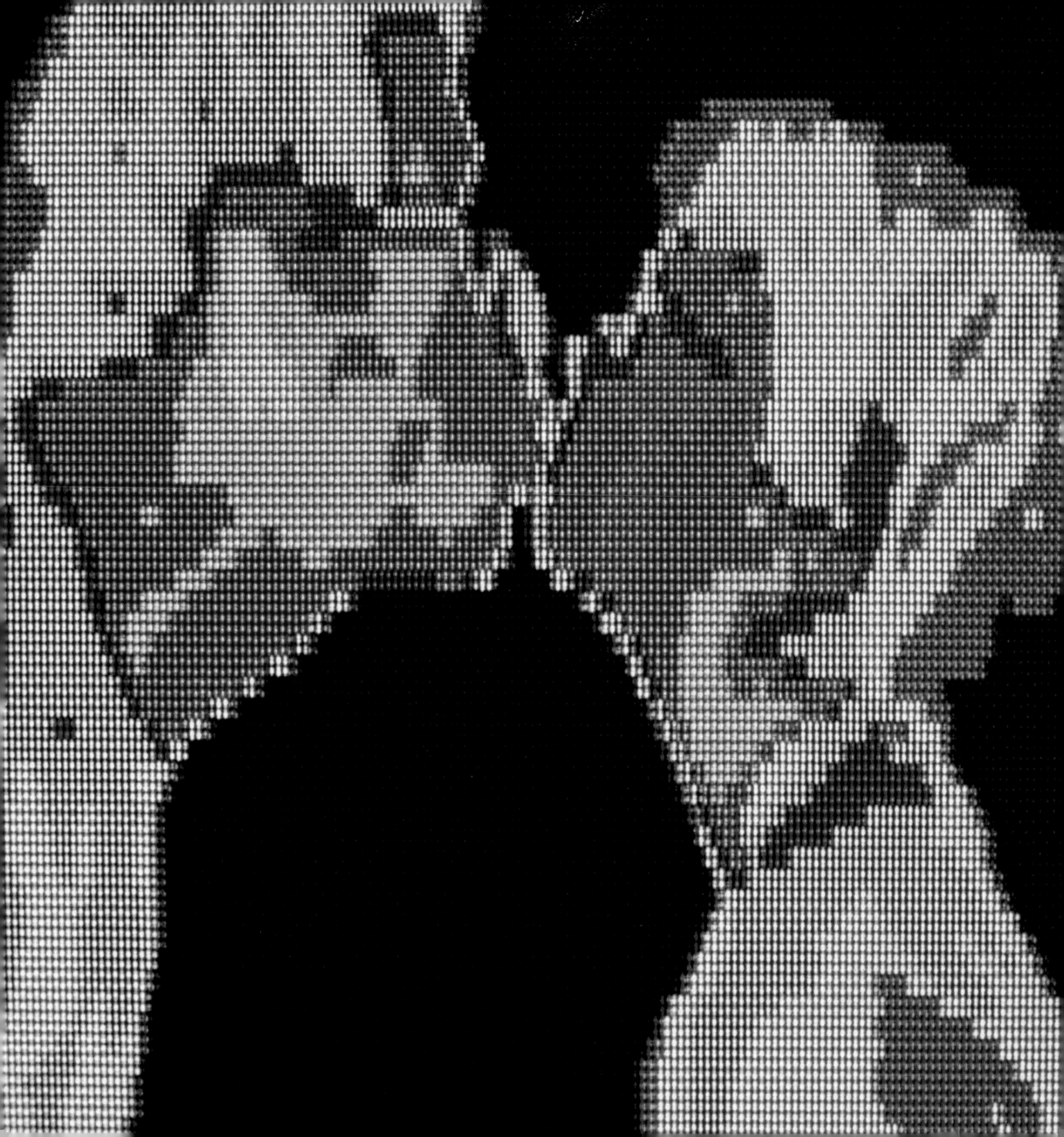

plants

71. How do nettles sting?

Stinging nettles are pretty plants with jagged green leaves that sting you if you brush against them. The Latin name is *Urtica dioica*, and the rash you can get when stung is called nettle rash, or urticaria. Each nettle plant produces either imperfect male or imperfect female flowers—in other words there are boy plants and girl plants—whereas the normal perfect flower has both male and female "parts."

The nettle stings are fine sharp hairs on the underside of the leaves and around the stem. They are loaded with formic acid, the same acid received from ant bites and wasp stings. When you touch the hairs, their sharp tips penetrate your skin like syringes and inject the acid. Some people hardly notice, while others are severely affected.

The acid of the sting can be neutralized by basic materials, such as bicarbonate of soda or baking soda, which will help to relieve the itching. Another helpful remedy is the juice of dock leaves. Dock (*Rumex* species) often grows near nettles, and has big, flat dark green leaves. Rub a dock leaf vigorously on the nettle rash and some of the juice will get in there and should alleviate the sting.

DID YOU KNOW?

Apparently the best way to apply dock is to twist the leaf up until the juice actually drips from it—much more efficient than just rubbing.

Challenge: The skin on the palms of your hands is generally too thick for the hairs to penetrate, so you can grasp a nettle firmly and pull it out of the ground. With luck you will not get stung at all, and you will certainly impress anyone who witnesses the deed.

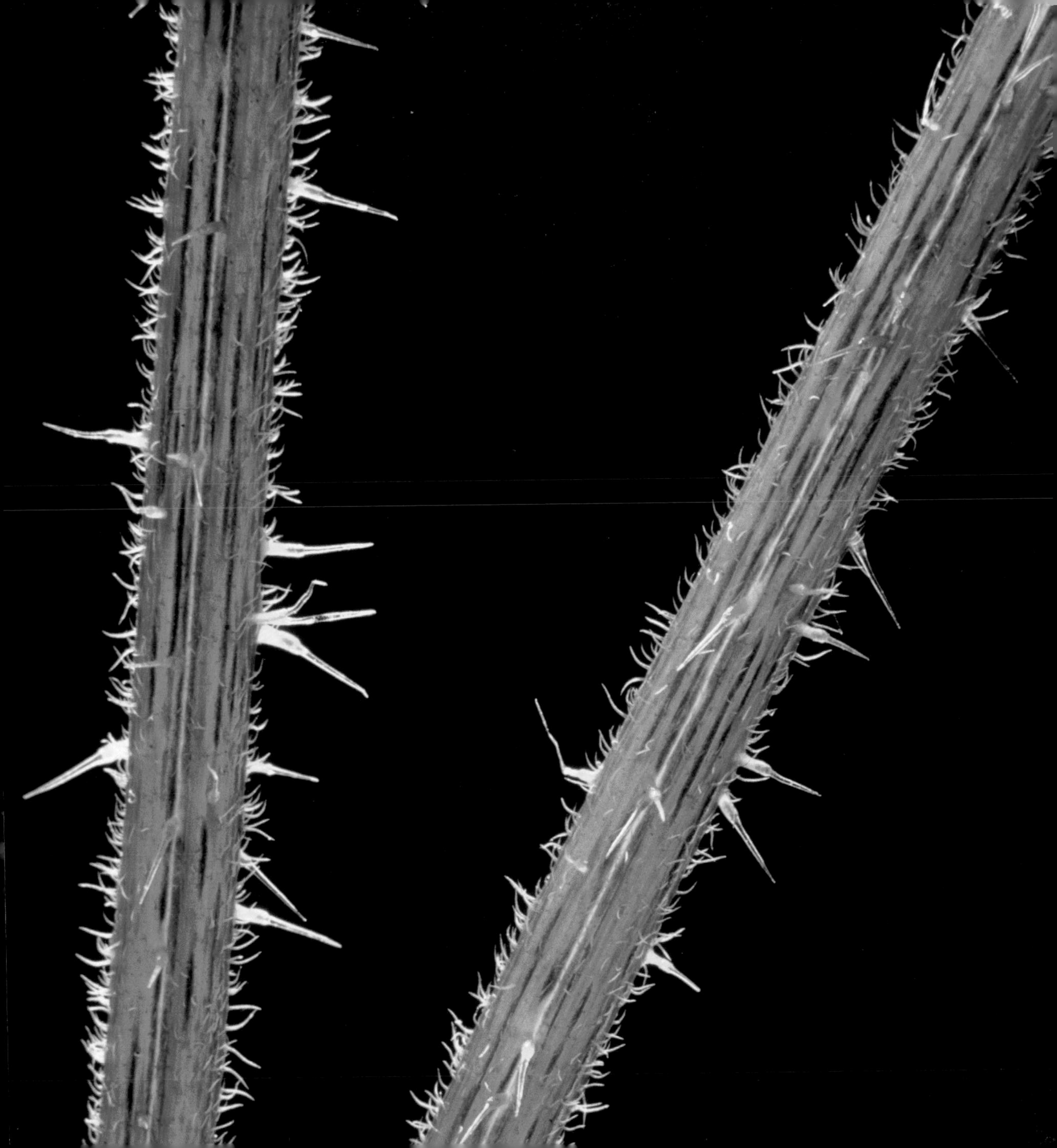

72. What are plants made of?

Some plants are puny, but trees can grow many dozens of feet high and live for hundreds of years. How do they build up such tremendous bulk and strength?

Trees and other plants get water from the soil, along with mineral nutrients they need to maintain their biochemistry through their roots. Most of their "food," however, comes from the air. Leaves have large numbers of tiny holes called stomata, through which they breathe. In particular, they take in carbon dioxide from the air and use it as "food."

Most plant leaves are green because they contain chlorophyll, a compound that absorbs sunlight and converts some of its energy into chemical energy. This energy is used to drive photosynthesis—a series of chemical reactions that turn carbon dioxide and water into sugars. A useful byproduct of photosynthesis is oxygen, which is breathed out through those stomata on the leaves. Without photosynthesis there would be no oxygen for animals to breathe and no plants for them to eat.

The sugars join together to make polysaccharides such as starch, which we eat in potatoes, cereals and pasta. Complex polysaccharides are good food, because our guts break them down over a long period of time to release the sugars needed by our muscles.

In plants the polysaccharides are slowly converted into cellulose and pectin, which are tougher than starch and form the basic building material for the cell walls and stems. Grass and other small plants are largely cellulose—the most abundant organic material on Earth.

For big plants, including trees, cellulose is not strong enough to stand up against the wind and weather, but over time further reactions take place to form lignin, a woody fibrous material that is hard and inert. A mixture of cellulose and lignin is the dietary fiber that we are often urged to eat.

So the short answer is that plants build themselves up by causing water to react with carbon dioxide. In other words, they are made of water and solid air.

Challenge: Pick some leaves, still on a twig, and put them under water in a jug or bowl on a windowsill in the sunshine. Do bubbles appear on the leaves? What are the bubbles made of?

73. What are roots for?

Virtually all land-based plants have roots, for two main reasons. The first is to hold the plant up. All green plants need air to breathe and light for photosynthesis (see page 153). This means they need to spread out their leaves and hold them as high as possible in order to avoid being shaded by other plants.

Tall trees are almost like sails—they catch the wind and are subjected to tremendous sideways forces. The leaves can twist away from the wind, the branches can move to lessen the force and even the trunk can bend. However, all the force must be taken by the roots, which anchor the tree in the ground and prevent it from falling over.

The same applies to smaller plants, such as hollyhocks, cabbages and even grass, although most of the stems are short, and the plants are rarely uprooted. A more serious problem is catastrophic bending of the stems. Torrential rain and high wind can cause terrible damage to such crops as wheat by simply knocking the plants over.

The second function of roots is to absorb water from the soil and pass it up to the plant. All plants need water, because all the chemical processes inside a plant happen in aqueous solution, and water is also food for the plant (see page 153).

Plants breathe water vapor out through the stomata on their leaves, and need to keep the balance by taking more in through their roots. Roots tend to start thick but branch rapidly, and are festooned with long fine tendrils or root hairs. These have an enormous surface area, so that the roots have access to the maximum amount of water, which (with some mineral nutrients) is drawn into the roots and pushed up the plant (see page 157) so that it eventually reaches all the leaves.

Leguminous plants, including peas and beans, have evolved an additional trick: they grow nodules on their roots to fix nitrogen. These nodules (opposite) contain bacteria—fed by the plant—that take nitrogen gas from the air and convert it into ammonia, which the plant uses to build up its protein. All plants need nitrogen, but generally they have to get it from fertilizer or from decaying plant matter such as compost or manure.

74. How do you make a blue carnation?

Simple. Take a white carnation, cut the stem on a diagonal (to increase the surface area of the cut) and stand the flower in ink or blue food coloring diluted with an equal volume of water.

Plants need to take in water through their roots, since they continually lose it through their leaves; if they don't take in enough water they shrivel up and die. They also need the water in the leaves to photosynthesize: to convert carbon dioxide and water into carbohydrate.

The sap that flows up the stem of a small plant or the trunk of a tree is mostly water, although it also carries a small amount of minerals from the soil. Plants need traces of minerals for their biochemistry, just as people do, and this is how they get them. Most of the plant's "food," however, comes from the air (see page 153).

The water is absorbed from the soil into the root hairs by osmosis—the solution is more concentrated inside the root, and so fresh water from outside is drawn in to dilute the solution. Then the water is pushed up by "root pressure"; if you cut off a grape vine at the ground, the severed root will continue to ooze sap.

Most of the flow of water, however, is due to the pull from above. Water is continually lost from the leaves by transpiration through the stomata on the leaves, and this provides negative pressure that sucks the water up from the ground. In addition, the water travels in narrow tubes, which provide capillary action—that is, water is always drawn up narrow tubes as long as it wets the walls.

Cut a tree down at ground level and stick the trunk in a bucket, and it will continue to "drink" water for a considerable time. This is why it is always a good idea to put cut flowers in water, because they will continue to transpire and pull water in through the stems.

From this photograph you can see that the water goes to the outer edges of the flower petals first, which suggests that the stomata are most active there.

Challenge: Can you make a two-color carnation—say red on the left and blue on the right?

75. What are flowers for?

In Victorian times the answer was simple. God had made human beings in His image to have dominion over the beasts of the field and so on, and He had created flowers to be pleasing to Man, or more often to Woman. There is, however, a real answer.

All living things reproduce. They have to; it is indeed one of the characteristics of living things. Some plants and animals reproduce asexually, by creating clones of themselves.

DID YOU KNOW?

Most dandelions are clones, and so are aphids.

Other plants and animals use sex for reproduction. Sex can be messy and cumbersome, but it has the advantage of providing the offspring with new combinations of genes. Clones are identical to their parents, but babies born as a result of sexual reproduction have genes from each parent. This provides variation, and means that the species may be better able to cope with changes in conditions, such as global warming, new diseases, swarms of locusts or other disasters.

Sexual reproduction depends on fertilization. Plants cannot walk around to find members of the opposite sex, so they have to be fertilized at a distance. They do this by pollination. The male parts of plants have stamens with anthers at the ends that produce pollen, which corresponds to animal sperm. The female parts have pollen receivers called pistils, which contain ovaries. The trick is to get the pollen from the anthers to the pistils.

Over the billions of years of evolution, one particularly cunning solution has been developed: plants and insects have come to depend on one another—this is called symbiosis. Bees in particular are attracted to flowers because they can collect nectar and pollen. They drink the nectar and collect the pollen on their hairy legs, then eventually return to the nest where they make honey. On the way, however, they visit more flowers of both sexes; so they collect pollen from the anthers and brush some of it off on the pistils. The flowers provide the bees and other insects with food, and in return get pollinated.

So plants have flowers to ensure their own future, because the flowers are signals to the insects that here is a supply of food, as bright and clear in the field as a neon sign for a restaurant.

76. Why do leaves have veins?

This is part of the underside of a leaf from a horse chestnut tree. The main job of the leaf is to soak up sunlight and convert carbon dioxide and water into carbohydrates—food for the whole of the tree (see page 153). The carbohydrates are made right across the surface of the leaf, and are then transported along the tiny veins into the bigger veins and finally into the biggest central vein to be carried out of the leaf. We have similar networks of veins all over our bodies, carrying the deoxygenated blood back toward the heart.

This pattern of branching veins is typical of flowering plants, or dicotyledons. By contrast, monocotyledons have narrow leaves and parallel veins.

Challenge: Can you find leaves from monocotyledons and dicotyledons?

The photographs on the right show the top and bottom of a giant water lily leaf, first grown in Britain in 1837 by Joseph Paxton (1801–65), head gardener at Chatsworth House in Derbyshire. When he realized that each leaf was almost 7 feet (2 m) across, he built a special pool for the plant with water wheels to keep the water moving. He named the lily after Queen Victoria, and presented her with a bud.

Paxton was astonished by the veins underneath the leaf. According to legend, he was inspired by their structure to design a modular greenhouse with a framework of cast iron and a glass roof. He went on to use an expanded version for the architecture of the Crystal Palace (see page 59).

77. Are seeds alive?

Biologists have always argued about what it means to be alive. Clearly there is a difference between a lump of rock and a squirrel, but what about a hibernating squirrel? What about viruses?

DID YOU KNOW?

One simple idea is there that are half a dozen characteristic signs of life: movement, respiring, sensitivity to external stimulus, eating, excreting and reproduction.

What about seeds?

Most animals obviously move, whether by bounding across a plain like antelope, or slithering through the soil like earthworms. Plants move too, often opening flowers and turning toward the sun.

Plants transpire (see page 153), mammals breathe and fish have gills. Both plants and animals are sensitive to changes in light and temperature; they take in "food" and water and excrete waste products; they start small and grow during their lifetimes, and they reproduce in various ways to make new generations.

But seeds don't obviously do any of these things. They can lie around for months, years or even centuries without showing any of the signs of life. Then, given a little moisture and a little warmth, they can spring into action and grow into new plants. Are they alive or not?

This photograph shows four wheat seeds planted on successive days. Within the first day water soaks in and the seed swells.

Challenge: Try growing your own seeds like this and watching them day by day.

Seeds provide the new shoots with the right starting cells to grow into a complete plant, and enough energy to reach down into the soil and up into the air. Once they are above ground, the leaves can start the process of photosynthesis (see page 153). Meanwhile the roots below develop fine hairs to take in the increasing amounts of water needed by the growing plant (see page 154).

Once the plant is independent the seed's job is done—the plant has successfully reproduced. The dry seed is neither alive nor dead but dormant. Give it water and warmth and it will soon show all the signs of life.

78. How do plants spread?

Plants grow from seeds and spread themselves by making new seeds. Because they are rooted to the spot, they can't go and sow them far and wide, and yet simply dropping them may not be successful, because it means that all the offspring have to grow in the same place as the parent and compete for water and light. Therefore, just as plants have evolved cunning strategies for sex (see page 158), they have also evolved cunning strategies for broadcasting their seeds.

Coconuts float, and coconut palms that grow by streams or the sea can spread their seeds by dropping the coconuts into the water; they sail away and may germinate elsewhere.

Cherries are immensely popular with blackbirds, which sometimes eat them on the tree, but often take them away to a safer place. Then they drop the cherry pits well away from the parent tree. Strawberries are also treats for birds, and their pips are so tiny that they get eaten with the fruit, and then deposited elsewhere in the bird's excrement—surrounded by their own nuggets of fertilizer.

Squirrels collect hundreds of nuts and need to store some for the winter when they want a bit of food but don't want to come out of hibernation for long. So they bury caches of nuts all over the place, but they often forget where some of them are, and so leave seeds neatly buried in the ground, ready to grow.

We have a pair of crows that live near our house and like the walnuts that grow on a tree across the road. They can't peck through the shells, so they carry them off for demolition. Sometimes they stomp around on our roof with heavy hoofs, bashing the walnuts on the tiles—but that does not work well. Eventually, they remember a clever trick and fly straight up a few dozen feet, and then drop the walnut on the road. If the shell breaks they walk around pecking out the goodies inside, and if not they pick it up and repeat the process. But occasionally by mistake they drop the walnuts in our garden, where they get lost in the flowerbeds and grow.

Sycamore trees turn their seeds into mini-gyros that spin away when they fall, while thistles and dandelions have seeds as light as down, able to float away on the wind.

79. Are hollyhocks masochistic?

The best hollyhocks in our garden occupy a narrow sloping flowerbed between a concrete path and a small bricked terrace. Each year the hollyhocks seed themselves, and seedlings crop up around the parent plants. Why do they flourish only in tiny cracks between the bricks on the terrace?

They could grow in the same flowerbed as the parents. They could grow in another bed only a few feet away. They could even grow on the lawn. But no; the seedlings seem to flourish only between the bricks on the terrace. These bricks are cemented in, and so if there are cracks down to the earth below they are small and few.

We have various possible explanations. There may be too much competition in the flowerbed, so that the tiny hollyhocks lose out to weeds and other flowers. This seems unlikely, but it is possible. Certainly there is no competition in this tiny crack.

The gap between the bricks may be particularly well drained, which may suit a baby hollyhock. On the other hand the sloping flowerbed must be fairly well drained too.

Seedlings that begin to grow in one of the flowerbeds may be attacked by slugs or snails (of which there are many). Their roots may be nibbled off by mice, voles, centipedes or other animals, which cannot get at either the roots or the first growing tip of the plant when it is protected by the bricks.

Or, more simply, mice may eat the seeds that lie on the earth, but cannot reach those that fall down between cracks in the bricks.

Perhaps the mortar between the bricks makes an alkaline environment, which favors the hollyhock seedling.

In theory we could do a controlled experiment: collect hollyhock seeds and plant them in matching places where we can vary one thing at a time: in a sloping flowerbed, in a flat animal-free flowerbed, in a well-drained flowerpot that is kept weed-free but left out for snails, in an alkaline spot in a protected flowerpot, and so on. Keeping all these experiments under control would be extremely difficult. I prefer just to marvel at the tough little plants.

Challenge: Have you a solution to this curious question? Can you come up with a convincing explanation?

80. Why do they all lean to the right?

These are tomato seedlings, just a few days old. Planted in damp compost and kept in a warm place, they all leap out of the ground with enthusiasm and a need for food. In order to make food, they need light, preferably sunlight (see page 153), and as much of it as possible, which means that, ideally, the surface of every leaf will be at right angles to the sunbeams. Therefore, the plant seems to want to rotate so as to lean directly toward the sun, which ensures that the maximum amount of light falls on the surface of its leaves.

In reality, the plants do not want anything because they have no capacity for wanting. However, imagine a hundred seedlings that pointed in random directions. Those that pointed toward the light would make more food and therefore grow more vigorously than the others. They would have more fruit and produce more offspring, and after a few generations there would be many more pointing toward the light than pointing the wrong way.

Then, in a further refinement, some plants would by chance rotate during the day as the sun moved around, so that they always turned toward the light, and these would be favored over the plants without any tendency to turn toward the light. This ability to turn with the light is called phototropism, and it is one of those millions of tricks that living things have "learned" through the long process of evolution by natural selection.

The mechanism of turning is a clever piece of biochemistry. A hormone called auxin makes the cells in the stem elongate, but it is switched off by sunlight, which means that only the shaded part—away from the sun—gets longer. This makes the plant bend over toward the light, and as the sun moves around, a new slice of the stem is inhibited, and the plant continues to bend toward the light.

Challenge: Grow some seedlings—tomatoes are easy—and try shielding the tray with dark paper or cardboard so that they get light from only one direction. Do they all bend that way? Will a table lamp do the trick? What happens if you change the direction of the light halfway through the day? Do they move, and if so, how long do they take to respond?

81. How alike are two peas in a pod?

When hunter-gatherers settled down and invented agriculture around 10,000 years ago, they learned how to grow crops that would yield a surplus of edible fruit. Wild grasses were developed into wheat and other cereals, and some leguminous plants became peas and beans. The fruit are the seeds for the next generation. The agricultural trick is to breed varieties that make many more fruit than is needed for next year's plants, and then to eat the surplus.

Peas have been available in many varieties for at least 150 years, which was useful for an Austrian monk called Gregor Mendel (1822–84), who was keen to investigate the consequences of sexual reproduction. He wanted to use mice, but his abbot disapproved of deliberately encouraging sex within the monastery at Brno; so Mendel used peas instead, and luckily the abbot failed to realize that plants have sex too.

Between 1856 and 1863 Mendel cultivated nearly 30,000 pea plants, carefully interbreeding varieties—wrinkled with smooth, tall with short, and so on—and laboriously worked out the basic rules of genetics and inheritance.

When he crossed tall and short, he found that all the first-generation children were tall, but when he interbred them, the second-generation children had one short plant in every four. He suggested there were things called genes that came in pairs and decided these characteristics—call them T for Tall and S for Short. One gene came from each parent; so the first generation would all have TS, and he said that clearly T was dominant, while S was recessive, because all the plants were tall.

In the second generation, however, with one gene from each parent, the possibilities are TT, TS, ST and SS, which means that even though S is recessive, one in four plants will have only SS and must therefore be short.

Unfortunately, Mendel's painstaking work was ignored for 30 years. Charles Darwin (1809–82) would have been delighted by it, because Mendel had answered a whole lot of the questions that Darwin raised in his book *On the Origin of Species* (1859).

The peas in a pod are not clones. Because they are the result of sexual reproduction, they have various combinations of genes and will grow into plants with differing features—or they would have if we had not eaten them for lunch.

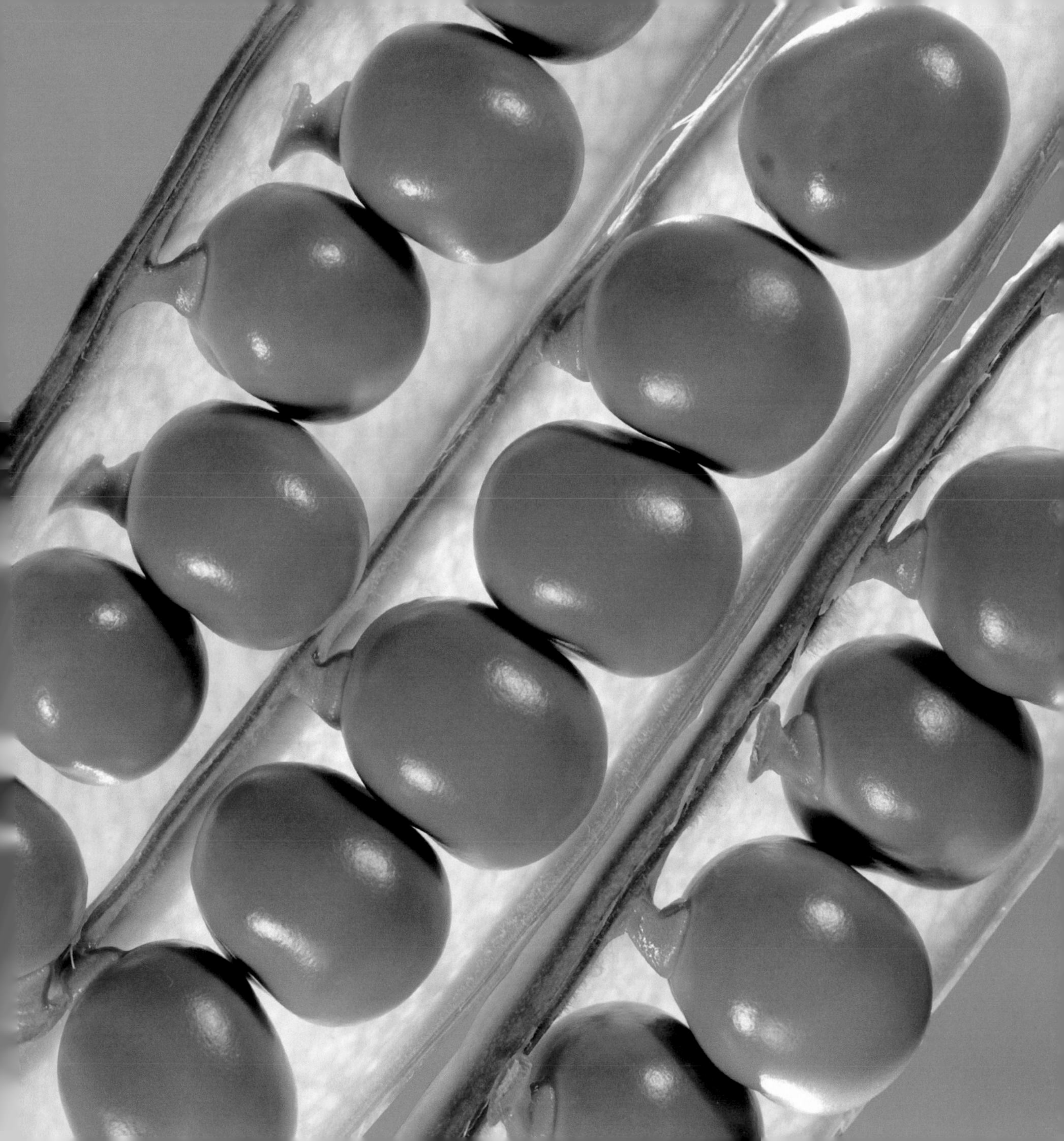

82. Who eats leaves?

The short answer is most land animals. All animals have to eat, and although there are plenty of carnivores—lions, spiders, sharks—there are even more herbivores—cows, sheep, chickens and so on—and herbivores eat leaves. Some herbivores eat shoots (giant pandas); some eat dead wood (termites); some eat seeds and nuts (birds and squirrels), but most eat leaves. There have to be many more herbivores than carnivores, otherwise the carnivores would run out of food.

There are also omnivores, who eat everything, or at least eat both animals and plants. Bears are omnivores, and so are people, although strict vegetarians are herbivorous.

Animals need to eat to get carbohydrates for energy and protein for muscle-building and repair. They also need small amounts of minerals and vitamins to keep their biochemistry going. Leaves are a useful source of carbohydrates, although digesting them can be tricky. We can't extract the goodness from the cellulose in grass, but cows can because they have several stomachs, and so the grass is able to ferment and break down inside. However, we can eat spinach, lettuce, cabbage and many other leaves, plus seeds of such grasses as wheat and barley, as long as they are ground up and turned into bread or beer. The problem with eating leaves alone is that they carry almost no protein, and therefore vegetarians—and most herbivores generally—need a source of protein, such as seeds or nuts.

Animals that eat only leaves, for example cows and elephants, have to eat almost continuously in order to get enough nourishment. In the case of elephants, most of those leaves go straight through the animal almost undigested.

Birds have an additional problem. Flying consumes a great deal of energy, and so they can't afford to eat indigestible low-calorie leaves. Therefore, there are no leaf-eating birds. Some birds are omnivores, and will eat worms, grubs, aphids and other birds' eggs, while other birds rely on seeds and nuts.

The leaf opposite has probably been eaten by a caterpillar, which has a short lifetime and exists mainly to become a butterfly or moth. Caterpillars spend most of their time munching away at leaves to build up stores of energy for the next stage in their life cycle.

83. Do vegetables have protein?

Most plants are made mainly of carbohydrates—cellulose for the stems and starchy or sugary stuff for the seeds and fruit. Some plants, however, make a lot of proteins, and that is why vegetarians are fond of legumes.

Legumes are the seeds of peas, beans, lentils and similar plants. These plants have developed a cunning trick. Nodules that harbor particular bacteria grow on their roots; these nodules can fix nitrogen (see page 154). This means that the plants are able to make amino acids, which all have nitrogen in them, without relying on the soil, and from the amino acids they can build up proteins.

Animals need proteins and have to eat them, or at least the amino acids, in order to build up the proteins they need. Legumes provide many of the proteins we need and the building blocks for some others.

The most prolific legume in the world is soy. Soy beans are grown in vast quantities both in the Far East—think of soy sauce and tofu, which are made from soy—and also in the United States, where for hundreds of miles in such midwestern states as Indiana and Illinois you can drive past huge fields that alternate between soy beans and maize, or corn. The corn is turned into corn flakes and popcorn, while the soy is ground up for animal feed and all sorts of human feed as well. Soy is particularly useful, since it contains the amino acid methionine, which we need, whereas peanuts and lentils do not.

Even legumes cannot provide all the protein that we need, but they provide an excellent base and form a valuable part of any vegetarian diet.

DID YOU KNOW?

Eating vegetables is far more efficient than eating meat, because a good deal of the energy from the food is wasted each time it is eaten. If people eat plants the food is eaten only once, whereas if people eat animals the initial plant gets eaten twice, and therefore there is twice as much waste. One way to provide more food for the increasing world population would be to persuade all the existing meat eaters to turn vegetarian, although this might be difficult.

84. How can you grow better vegetables?

Sandwiched between the rhubarb and the compost heap in the garden is my straw-bale urinal. I got the idea from the Centre for Alternative Technology at Machynlleth in Wales, and I use it for most of the year, apart from the middle of winter. It is also used by many male visitors and some women, though they tend to find it a bit prickly.

Challenge: Why not make one? Buy a bale of straw and stuff as much as you can into a plastic box, cut ends upward. Keep it tightly packed and you should get about a third of the bale into a normal stacker box.

When you pee on this, the urine disappears down between the stalks. It does not splash or smell, and it gradually decomposes in the middle of the bale. After a few months, depending on how many people use it, the straw darkens at the top, and in the middle has become brown and crumbly. Then you can tip the whole thing onto the compost heap and start again. The system works better if it does not get too wet; I put a lid over it in heavy rain.

DID YOU KNOW?

We urinate to get rid of waste products, especially surplus nitrogen. We need to eat nitrogen, in the form of protein, in order to build up and repair muscles and other tissues, but too much nitrogen is poisonous and we have to get rid of the excess.

Fish leave the excess nitrogen in the form of ammonia (see above), which is enormously soluble in water and is continually flushed out of their systems as they swim. We convert the extra nitrogen to urea, which is less toxic than ammonia, and send it off via the kidneys to the bladder.

Birds can't carry around all that water when they are flying, so they turn the nitrogen into a solid, uric acid, which is why bird excrement is mainly white.

Urine decomposes to make ammonia, which is readily absorbed and used by plants; plants need nitrogen as much as animals. Bear in mind that when you flush your urine away in a toilet you are wasting not only expensive drinking water but valuable fertilizer. By using a straw-bale urinal you will save water and be able to grow luxurious vegetables in your garden.

85. What's the oldest thing alive?

In general, big animals live longer than little ones. Bacteria survive only a few hours, insects a few days, mice a few years, cats perhaps 15 years, dogs up to 20, horses 30 and elephants around 50 years—although a few exceptional animals live much longer. Humans do rather better than other mammals, perhaps because of medicine, and some live to be more than 100 years old. Birds and reptiles can survive longer still; swans can live to be more than 100, and some tortoises and turtles live to more than 150 years—they are the Methuselahs of the animal world.

But even the reptiles are spring chickens compared to many trees. Oaks can live for many hundreds of years, junipers up to 2,000 years, giant sequoias for more than 3,000 years and bristlecone pines can live for 5,000 years (one cut down in 1964 was 4,950 years old).

Bristlecone pines grow in the desert in the southwest of the United States—southern California, Arizona and New Mexico. This photograph was taken near the Canyon de Cheilly, in Navajo country, where the canyon rims literally bristle with bristlecones. These soft pines, with coarse grain and leaves in bundles of five, look rather tired and decrepit, with the bark hanging off in strips, but don't be fooled—they are amazingly tough.

DID YOU KNOW?

According to the Bible, Methuselah, who was a great-great-great-grandson of Adam and Eve, had a son called Lamech at the age of 187, and lived on for a further 782 years, or as it says in Genesis 5:21, "And all the days of Methuselah were nine hundred sixty and nine years: and he died."

The easiest way to find the age of a tree is to count its rings. There is one ring for every year of growth (see page 113). You don't have to cut the tree down to do this. The scientists who measure the ages of trees—dendrochronologists—drill from the bark into the center and extract a narrow core of wood, just a few millimeters across. By examining this with a traveling microscope, they can see how many rings there are.

This technique has proved valuable in dating wooden things other than just trees, including the timbers of ships and buildings.

animals

86. Could dinosaurs swim?

This is the footprint of a dinosaur that once swam in the shallow waters at Whitby, on the northeast coast of England, or rather in the place where Whitby was built some time later. The animal left the impression of its foot in the soft mud, and the footprint filled up with silt that hardened over time. The whole thing became embedded in sedimentary rock and lay undisturbed in the cliffs for 180 million years. Then the sea began to wear away the cliff, and great chunks of rock came crashing upside-down onto the beach. The top of some of these rocks happen to be where the silt had settled on the mud, and the footprints of the dinosaurs are preserved in relief, standing up from the surface of the rock.

Clearly, this animal had three toes with claws, like a large bird. A short distance away you can see where it walked across the mud, but here the toe prints streak across the surface. The only sensible explanation seems to be that the animal was walking through a shallow lake or swamp, and then came to a deeper place where it could barely reach the bottom and began to swim, or at least float, scrabbling its feet along the bottom.

From the size and separation of these footprints, the paleontologists estimate that the animal stood about 2 feet (60 cm) high at the hip—like a turkey, perhaps—and was swimming in water a few feet deep. The first dinosaur bones were discovered in the early 1800s, and the word "dinosaur" was coined by English zoologist Richard Owen (1804–92) in 1841, but now dinosaurs seem to be cropping up all over the world, and we learn new things about them every year.

87. What are sea urchins?

This group of shells, picked up on a beach, once housed sea urchins. I took the photograph because I was intrigued by the intricate patterns of lumps and perforations, but I also wondered what sort of animals had lived inside. The short answer is that sea urchins are little things that lurk on some seashores, and spike your foot if you walk on them. Biologically, they are echinoderms and belong to the same family as starfish. Starfish have five "arms," and if you look carefully you can see that these sea urchin shells have five segments; in other words, they have fivefold symmetry. That means that if you look from the center along a radial line of perforations, and then circle around 72 degrees (360 ÷ 5), you will find an identical line of perforations.

You might expect fivefold symmetry to be rare in nature, for two reasons. First, pentagons and similar objects with fivefold symmetry do not tessellate (see page 114), that is, they do not pack together tidily, filling all of a space. Second, nature tends to work in pairs, producing living creatures with bilateral symmetry, in other words left and right halves that match. Human beings have matching left and right, ignoring minor differences; so do other mammals (dogs, cats, horses, etc.), birds, fish, insects and so on. None of them has five legs.

So echinoderms are rare in the natural world. Is it just a coincidence that five is one of the Fibonacci sequence? (See page 121.)

Sea urchins sit at the bottom of shallow water, holding onto rocks with little sucker feet, of which they have five pairs of rows. They are protected by this hard shell or test, and through the dozens of perforations come tiny hairs or tentacles, which are poisonous for protection and are also useful for catching drifting algae.

The spines that stick into your feet help the beast to move about, and between them are tiny pairs of tweezers that it uses for self-defense and grooming. The sea urchin chews its food with five jaws, known as Aristotle's lantern; it feeds on algae, kelp, dead fish, mussels and barnacles, and is in turn eaten by crabs, snails and people. It has a mouth underneath, an anus on top and no brain anywhere.

88. What use are seashells?

Seashells are the exoskeletons of marine molluscs or, in other words, the houses of animals that live in the sea. Some animals—cats, fish, people—are vertebrates; they have backbones and other bones that allow them keep their shape, and all the vital organs hang in bags around and between the bones, which means that these vital organs are poorly protected from the dangers of the world.

More cautious animals build themselves solid boxes to live in. These boxes or houses give them good protection, but are a hindrance to movement. Snails don't run fast and limpets are even slower. Nevertheless, these creatures manage to move quite effectively. Clams, for example, swim by "clapping" the two halves of their shells together.

The seashells in this photograph are made of calcium carbonate, chemically the same stuff as limestone and marble. The saltiness of seawater is due mainly to sodium chloride, but there are other chemicals dissolved in the water, including calcium and carbon dioxide, which is absorbed from the atmosphere and swallowed up by green plants (see page 153). Every marine mollusc spends part of its life eating plants, taking in calcium salts and converting them into calcium carbonate.

Seashells are built up gradually, year by year, and have growth rings, rather like trees (see page 178). By examining fossil shells, biologists can piece together information about the temperature and saltiness of the sea in the distant past.

DID YOU KNOW?

Ancient seashells get squashed together on the seafloor, compressed further by more layers on top, and eventually turn into limestone, which is why limestone often contains marine fossils.

Humans and other vertebrates turn small amounts of calcium salts into bones and teeth, but these marine molluscs build shells that weigh more than the rest of their bodies, and carbon dioxide accounts for nearly half the weight of every shell. So shells lock up large quantities of carbon dioxide and keep it out of the atmosphere. Millions of tons of carbon dioxide are buried at the bottom of the sea in old seashells, and without these helpful creatures we might long ago have succumbed to rampant global warming.

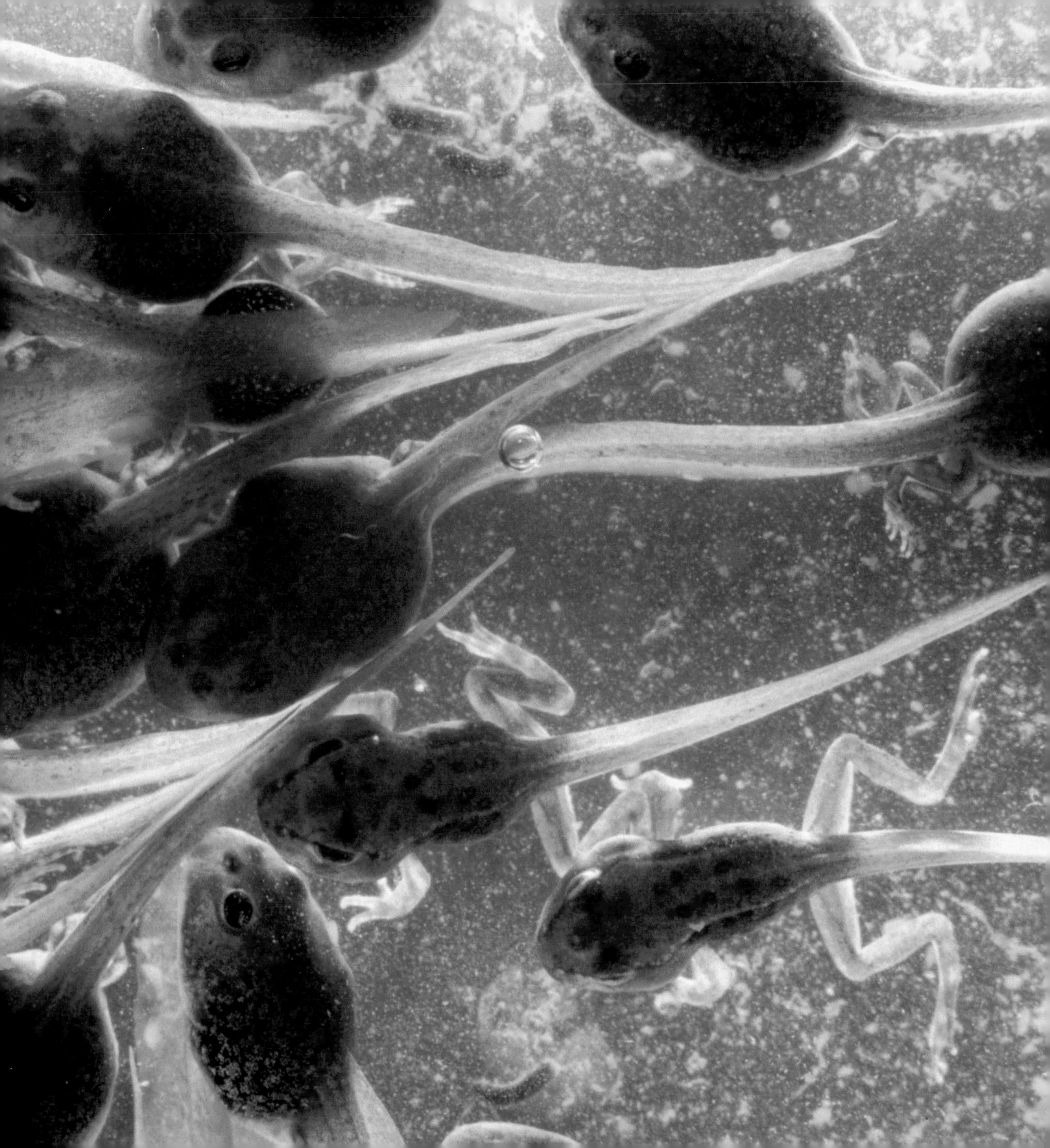

89. What are stem cells?

Our little garden pond is home to a few frogs. In early spring they begin mating and frogspawn starts to appear—larger and larger globs of translucent jelly, with a little black dot, the egg, in the middle of each capsule. The capsules separate, and suddenly the pond is alive with wriggling tadpoles, which at first are just a round head and a wiggling tail. Then they become lighter brown, and almost as if by magic they begin to grow legs.

The best thing about frogs is that you can watch them growing from a primitive dot to a fully grown adult. In the frogspawn the cells look very similar; to begin with there is no sign of legs or head. Then the head and tail appear, and much later the limbs. This raises the question: how do those cells decide to become legs? In the egg, all the cells were the same, yet a couple of weeks later those cells are no longer identical—they seem to have decided their destiny.

One of the challenges facing developmental biologists is to study and, if possible, make use of embryonic stem cells, which are cells that have not yet decided what they are going to grow into, and could give rise to any sort of cell in the body. If we did grow and harvest these stem cells we might be able to regrow broken nerves and therefore reverse paralysis. We might even be able to regrow severed fingers or arms, and stem cells may provide new ways of tackling genetic diseases.

DID YOU KNOW?

Human beings are in many ways similar to frogs. We start life, after conception, as just a few cells, all identical. Then a little later the cells differentiate to form the body; some become leg cells and some head cells; some become nose cells and some eye cells.

Each female frog can produce hundreds of pieces of frogspawn. If every cell developed into an adult frog the pond would overflow with frogs, and so, probably, would the garden. At least 99 percent of the offspring must die if there are to be roughly the same numbers of frogs in the garden next year. Most of these will be eaten either as spawn or as tadpoles by bigger frogs, fish, birds and other animals, including our cat, who eats frogspawn from time to time, and then thoughtfully throws it up on the lawn.

90. How do spiders catch flies?

Spiders caught my attention one autumn in Yorkshire, in the north of England, when the whole garden seemed to be full of beautiful webs, and I began to take photographs of them. I did some research and was amazed by their variety. The most common in my garden turned out to be, appropriately, the common garden spider, *Araneus diadematus*, which is brown with a large pale cross on its back. It spins orb webs between any convenient anchor points, such as projecting twigs of trees and bushes, fences or blades of grass.

DID YOU KNOW?

There are more than 35,000 species of spiders in the world.

Spinning a web takes about half an hour. Producing silk from a spinneret on its rump, the spider first makes links to the main anchor points, then perhaps 25 radial spokes from the center to further anchor points if it can find them. When there are enough spokes it goes around and around, spiraling outward from the center, to build the cross-ribs, all the time coating the silk with a thin layer of glue that sticks the threads together at crossing points and helps the web to be a good trap for flies. Altogether there may be 65 feet (20 m) of silk, with 1,000 junctions and a total weight of only half a milligram. The web needs frequent repairs, and the garden spider makes a new one every day, eating the old one to recycle the silk.

When the web is finished, some spiders retire out of sight, perhaps to a crevice nearby, holding onto a single strand that is connected to the center of the web. The female garden spider often does not hide, but hangs head down in the center of the web, holding onto the spokes with all eight feet. When a fly blunders into the web it gets stuck on the sticky strands and struggles to get free. This makes the web vibrate, and the vibrations alert the spider, which sprints up the right spoke to catch the fly.

DID YOU KNOW?

Spider silks are made of protein, manufactured by the spider, and the dragline silk used for the spokes is weight for weight stronger than steel. Genetic engineers are looking for ways to make dragline silk, which could have important uses in medical technology.

91. Why do bees make hexagons?

A colony of honeybees has one queen, a dozen or two male drones to mate with her and perhaps 50,000 female workers. Some years ago a swarm appeared in my garden. The new queen must have been looking for a good home, and on the way they stopped off for a rest at the end of a hedge. Luckily, our neighbor was a beekeeper, and he was happy to come around with his veil and gloves and take the swarm off to an empty hive. In order to move it he persuaded the bees into an old cardboard box, and while they were there they made a little section of honeycomb, which I later photographed.

The cell walls are made from wax, which the worker bees sweat from their abdomens in the form of scales after eating loads of honey and pollen. They build the cells on both sides of vertical walls, the inner ones for the queen to lay eggs in, and the outer ones for storing pollen and honey. They tilt the cells up slightly to prevent the honey from running out before the end of the cell can be capped with wax.

DID YOU KNOW?

Bees don't really make hexagons at all. They merely make cells into which they can just fit. These cells are essentially tubes; so when the bees pack the cells together as tightly as they can, the tubes pack like those on page 122. Six tubes fit neatly around one tube, and because the walls deform slightly from the pressure of their neighbors, the shape of each cell looks hexagonal.

Once they have extracted the honey, human beings use beeswax to make excellent candles, furniture polish, crayons, cosmetics, saddlesoap, and lubrication for sewing thread, drawer slides and didgeridoo mouthpieces—though this last one comes from stingless Australian bees.

Wasps make nests using paper, which they produce by chewing up cellulose from plants and spitting it out again as a sticky mass. The nests are constructed in holes in walls, or inside roof spaces, where they will remain dry. They can be big, holding hundreds or even thousands of wasps. Their construction is beautiful, and the individual cells appear hexagonal just like those of the bees.

RATATO
PROVE

92. How bright are rats?

Human beings have large brains—about three times as big as those of the apes, our nearest relatives. Humans are highly intelligent—we have created books and phones, music and bicycles, achievements way beyond the reach of any other animals.

Nevertheless, many animals do show intelligence. Apes can pile up boxes in order to reach bananas, and crows are capable of bending a piece of wire into a hook to pull food out of a tube. Some of the most studied of all the animals are rats.

This rat, a pet called Shniffles, was a brown hooded rat. Rats make good pets and also good laboratory animals, because they are easy to breed, cheap and easy to keep, and intelligent. They can be taught all sorts of things.

Psychologists generally use white rats, which are albino forms of brown rats, and have used them for years to investigate what they call "operant conditioning," which really means learning from the environment. For example, rats will find their way through a maze and then remember the correct path next time.

The famous behavioral psychologist B.F. Skinner (1904–90) did a great deal of work with rats. In particular, he invented what came to be called a "Skinner box" in which a rat could get food by pressing a lever. The idea was that the rat would run around inside the box and occasionally happen to press the lever. It would soon discover that this produced food, and would learn to associate the two things; so it would press the lever again and again to get more food.

DID YOU KNOW?

There are dozens of kinds of rat, but the best known are the brown rat and the black rat, both of which carry fleas that can spread the plague and other horrid diseases.

Then Skinner moved the goalposts and arranged the mechanism so that the rat got the food only after three presses, or after two presses within five seconds, and so on. The rats turned out to have considerable skill in working out the changing rules.

B.F. Skinner was said by some to be almost like a magician—forever pulling habits out of rats.

93. What is slug slime?

Gardeners hate slugs with a deadly loathing. Imagine a row of fresh green seedlings, poised to grow into tall, strong bean or tomato plants, yielding pounds of food. Along comes a slug in the night, and it mows down the entire row for a quick snack. Grrr.

Slugs are always in danger of drying out; they like moisture and appear in droves after rain. They are gastropods, meaning "stomach-walkers," — their organs are twisted back so that the stomach sits on top of a big muscular foot. They move by contracting the foot's muscles in waves. To help slide over the ground—and cling to vertical surfaces—they ooze slimy mucus from a gland in the front, where their toes should be.

Slug slime is clever stuff; slugs use it as a mating rope and a navigation aid. It can absorb a huge amount of water, which helps to protect the animals against dehydration. It tastes horrible, and so protects against some predators. It's sticky (one group of high-school researchers found that it makes good glue). And it's difficult to get off your hands—don't try washing until you have done your best to rub it off dry first. Technically, it is a glycoprotein polymer, rather like snot.

The two big tentacles on top of slugs' heads have light-sensitive "eyes," while the small ones in front deal with smell and touch. Slugs have rough tongues, or radulas, with minute teeth, or denticles, that tear holes in plants, and they breathe through an air hole, or pneumostome (visible in the photograph), although they also absorb oxygen directly through the skin.

They dig deep for shelter from cold and drought, emerging in spring to lay hundreds of eggs in damp soil or rotting leaves. Slugs are hermaphrodites and can fertilize their own eggs, but generally prefer to cuddle up to another slug. There can be a hundred or more in 10 square feet (1 sq. m) of garden, so getting rid of a few, or even dozens, does not help much. What you need is a family of hungry moles, hedgehogs, raccoons or other animals that eat slugs. Slugs can be poisoned with metaldehyde or methiocarb or common salt, and have a deadly enemy: a nematode parasite with the snappy name of *Phasmarhabditis hermaphrodita*, which infects the slug and kills it in a couple of weeks.

health

94. What happens where in your head?

During the 19th century, science really took off in Britain. Many clergymen, being educated, reasonably well paid and having time to spare, carried out observational and experimental work. In 1840 William Whewell (1794–1866) wrote: "We need … a name to describe a cultivator of science in general; I should incline to call him a scientist."

Nineteenth-century science was not all brilliant, phrenology being a case in point. The pioneer, Viennese doctor Franz Josef Gall (1758–1828), asserted that the brain had separate organs for benevolence, intelligence, murder and so on. Each was powerful in relation to its size, and because these organs defined the shape and size of the skull, you could find out about the owner's personality from the bumps on the outside. Soon the entire skull was mapped with these various organs.

In the United States, the idea was embraced by the Fowler brothers, Orson Squire and Lorenzo Niles. In 1863 Lorenzo emigrated to England to give lectures, examine believers for large fees and found the British Phrenological Association. What a pity that today's brain scans show that phrenology was utter nonsense.

DID YOU KNOW?

Whewell was Master of Trinity College, Cambridge, and once had the honor of showing Queen Victoria around the university. There were no proper sewers then and everything discharged straight into the River Cam. Recoiling at the smell, the Queen asked "Tell me, Master, what are those pieces of paper floating in the river?" Quick as a flash, Whewell replied "Those, Ma'am, are notices saying that bathing is forbidden."

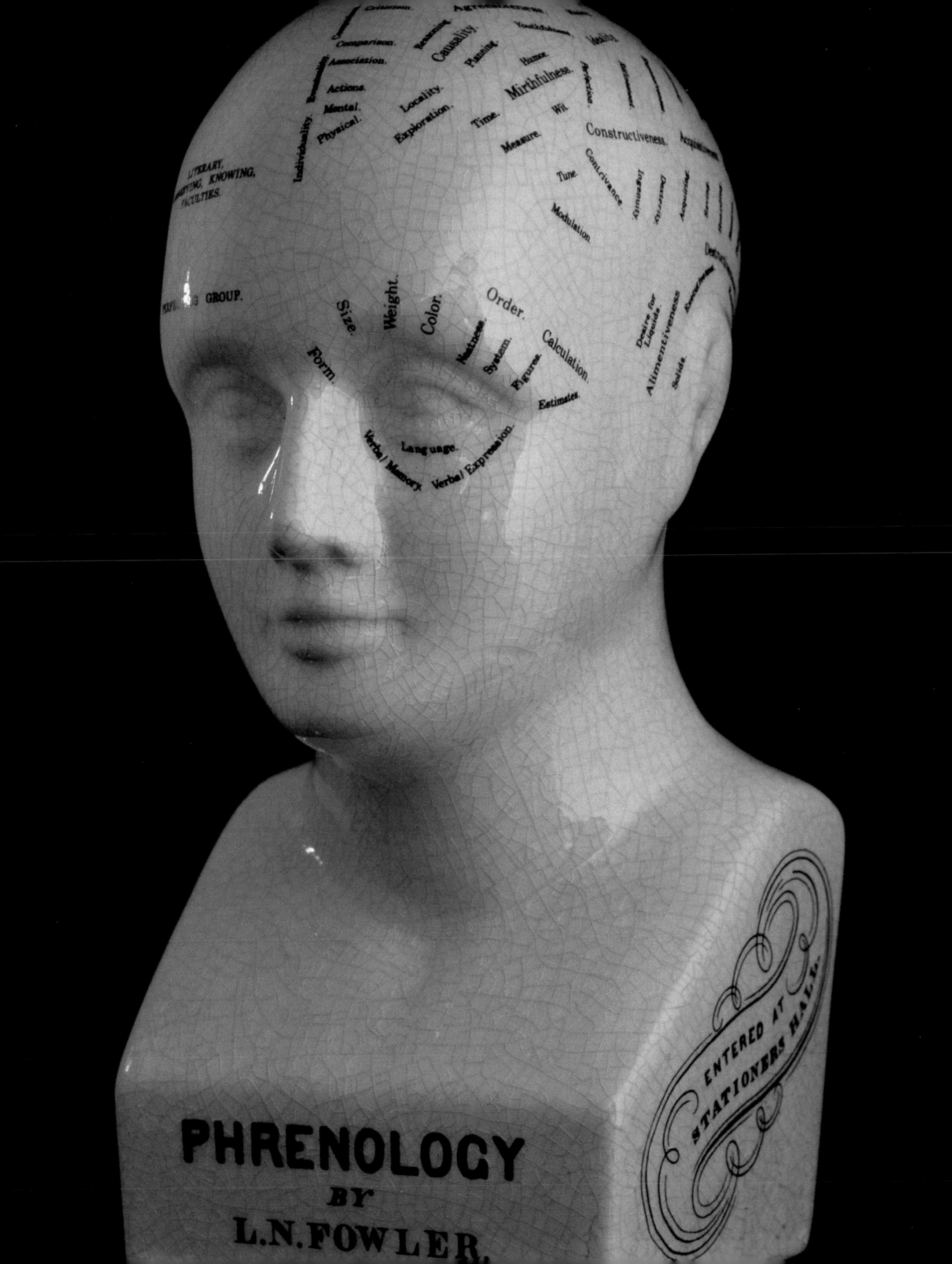
Comparison.
Association.
Actions.
Mental.
Physical.
Individuality.
Causality.
Planning.
Locality.
Exploration.
Time.
Measure.
Wit.
Mirthfulness.
Constructiveness.
Tune.
Modulation.
Contrivance.
Size.
Weight.
Color.
Order.
Form.
Neatness.
System.
Figures.
Calculation.
Estimates.
Language.
Verbal Memory.
Verbal Expression.
Desire for Liquids.
Alimentiveness.
Solids.
PHRENOLOGY
BY
L.N.FOWLER.
ENTERED AT
STATIONERS HALL.

95. Why do some plants affect the mind?

The mind is a construction of the brain, and some chemicals affect the mind because they affect the brain. For example, alcohol, painkillers and anesthetics all affect the brain, and therefore the way we interact with the world around us. Some of these brain-changing chemicals are synthetic, but many are natural—plants have a bewildering array of active ingredients.

This is a leaf of *Cannabis sativa*, which people have been using for many thousands of years and will probably continue to use for thousands of years to come, no matter how hard authorities try to prevent them. From the stems come fibers of hemp, which is splendid material for clothing and ropes, while the leaves and flowers are easily turned into potions that can be used for medical or recreational purposes. In various forms, cannabis (dope, grass, weed, hash, marijuana) can be eaten, drunk or smoked. Users report every sensation from improved sexiness and euphoria to foggy thinking and paranoia. Laboratory tests have shown that it is bad for short-term memory but has little effect on long-term memory, and medically it has been shown to be good for pain relief, especially for people with multiple sclerosis.

DID YOU KNOW?

The brain is like an enormous tangle of electrical wires, except that each wire is a living cell or neuron that communicates with hundreds or even thousands of other neurons. The tiny gaps between neurons are called synapses, and messages are passed across the synapses by chemicals called neurotransmitters, which leave one neuron and jump across the gap to a special receptor on the next.

The active chemical in cannabis is tetrahydrocannabinol, or THC for short. THC acts by affecting the neurotransmitter receptors in the brain.

Drugs that affect our state of mind do so by interfering with these neurotransmitters. For example, antidepressants such as Prozac work by increasing the amount of a neurotransmitter called serotonin, while cocaine works by increasing the amount of dopamine. Other drugs mimic the normal transmitters they "pretend" to be transmitters and attach themselves to the receptors, thus either sending false signals or blocking real ones. THC seems to be rather special; it has its own specialized receptors, almost as though the brain is expecting it to come along ...

96. Is there a pill for every ill?

Throughout human history people have felt sorry for themselves when they were ill, and therefore there has always been employment for doctors and suppliers of remedies. Until the last hundred years or so the only useful medicines came from plants. For example, the bark of the cinchona tree turned out to be effective against malaria, and chewing willow bark was good for easing the pain of a headache—but no one knew why.

One of the first to try innovative chemical treatment was the 16th-century alchemist Theophrastus Philippus Aureolus Bombastus von Hohenheim, also known as Paracelsus (1493–1541). He treated patients with poisonous metals including arsenic and mercury, and probably killed more than he cured, but he had the idea that the disease was a living entity, and therefore susceptible to chemical attack rather than an affliction of the gods. This idea eventually bore fruit in 1909 with the invention of the first effective treatment for syphilis, Salvarsan, which contained arsenic. In fact, it was the 606th arsenic compound tested by German bacteriologist Paul Ehrlich (1854–1915). Salvarsan was hailed as a "magic bullet," which killed the disease but not the patient.

Meanwhile, the most popular drug of all time had already been made in Germany. The active ingredient in willow bark is salicylic acid, but it makes you sick. In 1897 Arthur Eichengrun at the chemical company Bayer modified the compound by treating it with acetic acid (the main constituent of vinegar) to make acetylsalicylic acid (ASA), which Bayer marketed as Aspirin. ASA has turned out to have an astonishing range of properties, reducing fever and pain, thinning blood, and so reducing the risk of heart attack, stroke and some forms of cancer.

Paracelsus's other idea, that there should be one specific pill for every ill, was wide of the mark in the case of ASA, and even wider in the case of penicillin, discovered by Alexander Fleming (1881–1955) in 1928. Although making the stuff was difficult, it was extraordinarily effective against bacterial infections of many kinds.

Today, knowledge of biochemistry and genetics has advanced to such an extent that there are real possibilities of developing a specific pill for every ill, and vaccines for many, but nature is at least as clever as we are, and we are still unable to do much to relieve the common cold.

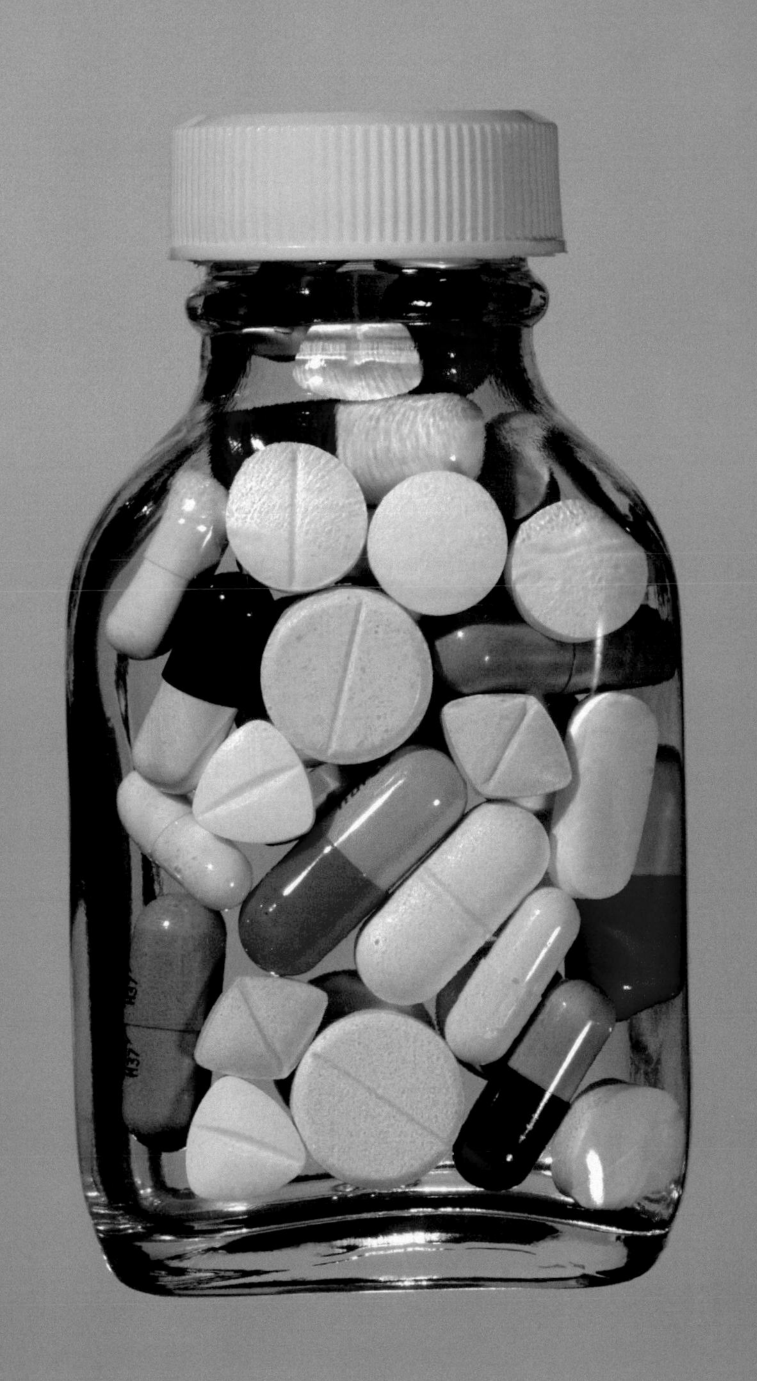

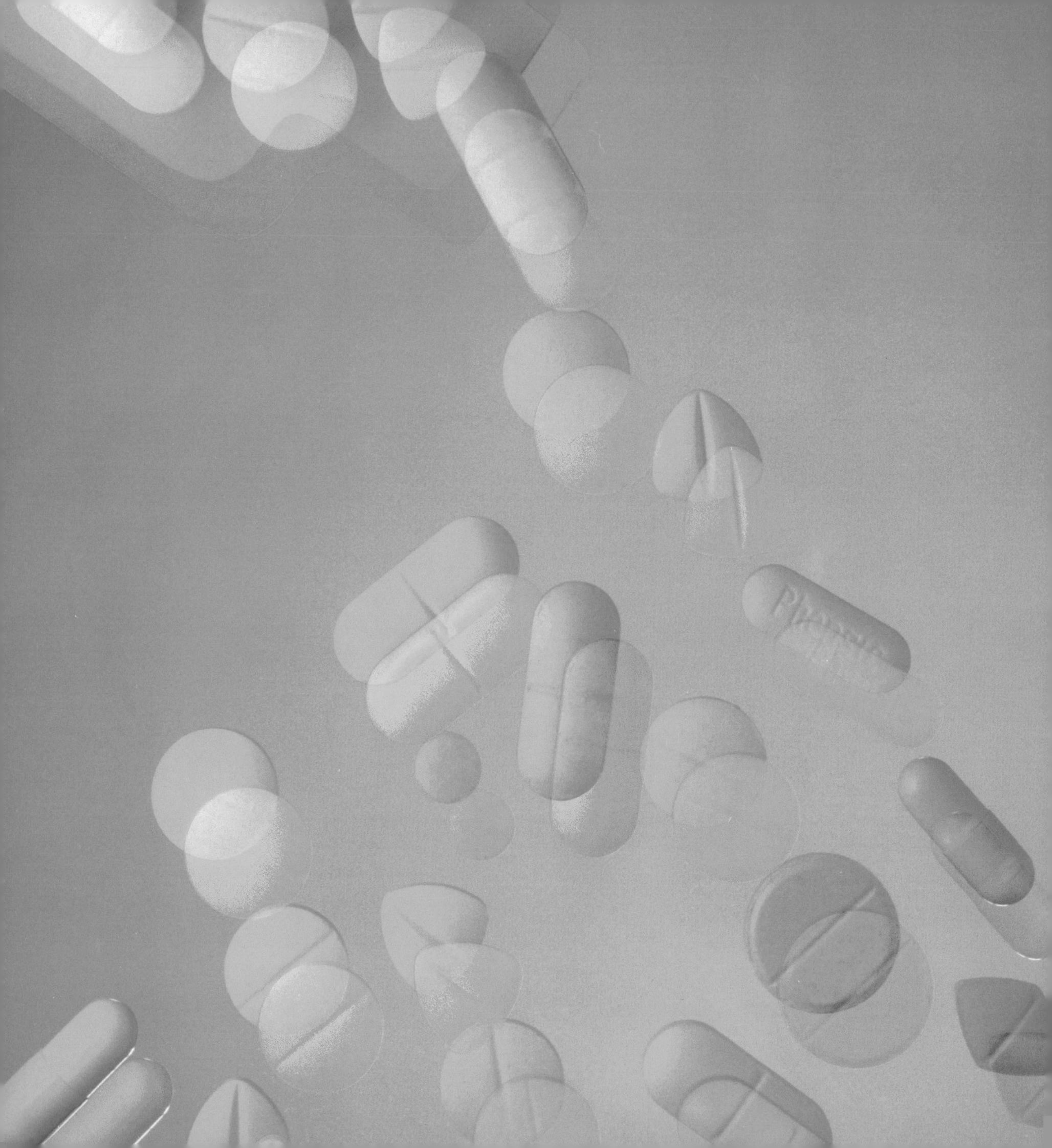

97. What are pills made of?

When I visited a pharmaceutical factory recently, the first thing I learned was that to the people who make them these things are not pills at all, but tablets. Pills were made behind the counter by old-fashioned chemists.

DID YOU KNOW?

The name depends on the shape. "Tablets" are generally round, although some are triangular, square, etc. "Capsules" are long, shiny cylinders of gelatin, often colored, and made in two sections, joined around the waist. The shape makes the capsules easy to swallow, and the container hides the taste of the drug inside, which is released when the gelatin dissolves in the stomach. "Caplets" are tablets made roughly the shape of capsules, to ease swallowing, but without the gelatin container.

Up to three-quarters of each tablet is the drug—Aspirin, acetaminophen or whatever. The rest is a mixture of additives that simplify manufacture and improve effectiveness.

For each tablet, the drug starts as a powder, but powders are difficult to handle—they don't pour or mix easily. So how do you turn the powder into a solid, stable tablet? First, add about 30 percent of a filler such as lactose (a form of sugar): if the correct dose of drug is 100 mg, add 30 mg of lactose. Then, to tame the sticky powder, add perhaps 4 mg of a binder, usually modified cellulose. The mixture goes into a giant food processor and comes out as granules, which after drying and sifting are like table salt—much easier to handle than powder.

The granules could be compressed into tablets, but these would be slow to dissolve in the patient, so add perhaps 10 mg of what they call a disintegrant like modified starch, which will absorb water, swell and break up the tablet in the stomach. Finally, add a few milligrams of a lubricant such as talc, which makes the granules flow more easily and compress more cleanly.

The slippery granules are compressed in stainless steel dies at high pressure (equal to several thousand atmospheres), and out pour the products—250,000 tablets an hour—from a single machine. Finally, some tablets are coated with a film of modified cellulose to mask the taste or delay release of the drug.

98. Why bother with vaccinations?

Many people prefer to avoid injections, so why do doctors like vaccinations? The short answer is that five seconds of discomfort is much better than days or even weeks of a serious illness. Vaccination prevents disease.

The idea of deliberately giving someone a disease in a mild way to prevent a more serious attack seems to have originated in India, and was eventually brought to Britain in 1721 by Lady Mary Wortley Montague (1689–1762). In those days, primitive inoculation against smallpox was unpleasant and dangerous; many victims actually died of the treatment. One boy who hated it was Edward Jenner (1749–1823), who became a country doctor in Gloucestershire, was elected to the Royal Society for a paper he wrote about cuckoos, and invented vaccination —using cowpox.

In 1879 the French scientist Louis Pasteur (1822–95), investigating a different disease in chickens, discovered almost by mistake how to make "attenuated" germs—that is, a mild form, rather similar to the cowpox.

DID YOU KNOW?

Jenner had an eye for a pretty girl and noticed that milkmaids had no pockmarks; they never seemed to catch smallpox, even though most people did. He realized this was because they caught cowpox from cows. Cowpox made them ill for only a day or two, but it gave them immunity to the deadly smallpox. In 1796 Jenner deliberately infected a young boy, James Phipps, with cowpox, and when he had recovered tried to infect him with smallpox. He failed, and Phipps lived to a ripe old age without ever catching smallpox. When Jenner wrote up his conclusions, he called his procedure vaccination, after *vacca*, the Latin for cow. He was not the first to try this process, but people listened to him because he was a Fellow of the Royal Society. Vaccination was taken up on a worldwide scale, and now the World Health Organization records no smallpox in any population on Earth.

Today, we vaccinate babies and children against a range of diseases that were once common. Measles, for example, used to be accepted as something that most children caught during their school years—I remember having it myself—but it can cause brain inflammation, pneumonia and disability. It can even be fatal. Now children are protected by vaccinations so that they are spared these possible disasters.

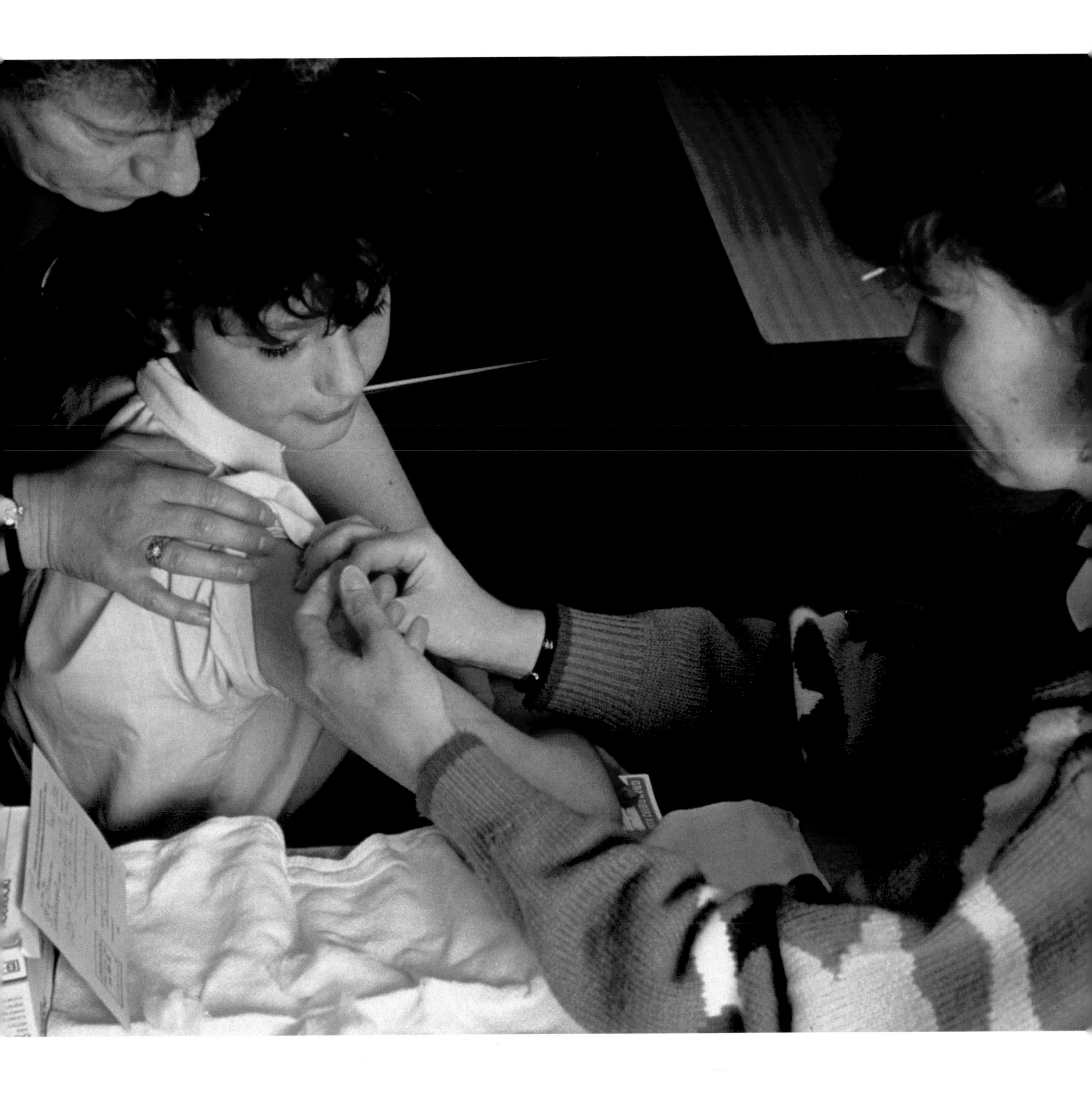

99. How do diseases spread?

Until the 1840s no one had a clear idea about how people caught diseases. Malaria was thought to come from "bad air" (which is what "malaria" means). It was not until the end of the 19th century that British physician Ronald Ross (1857–1932) worked out the truth about mosquitoes.

DID YOU KNOW?

Ross wrote to the Swedish government and asked for a Nobel Prize. They replied that they would choose who deserved to win Nobel prizes—and in 1902 they gave him one.

The process of tracking diseases is called epidemiology, and among the first epidemiologists were William Budd (1811–80) and John Snow (1813–58). Budd was a British army doctor who caught typhoid while working on a hospital ship moored at Greenwich, England. Luckily he survived, and was sent to recover at home in the village of North Tawton in Devon.

He was much concerned and professionally interested when typhoid came to the village in 1839, and family after family was affected. Because he knew the place and the people well, he was able to establish precisely the disease's path from house to house, and he became convinced that it spread through the water supply. One simple piece of evidence was that a family living by a stream from which they took their drinking water contracted the disease a few days after a family who lived upstream. Later he moved to Bristol, and there was able to prove that an outbreak of cholera was also transmitted by contamination of the drinking water.

Meanwhile, a London doctor, John Snow, came to the same conclusion about the cholera outbreak of 1849, and wrote a paper explaining his theory. Few people believed him, but then there was another outbreak in 1854, and several hundred people died in the area around his practice in Soho. He showed that the water from the Broad Street pump was contaminated, and in a dramatic move had the handle removed from the pump so that no one else could drink the water. Within days the outbreak died down.

Today's epidemiologists work in public health laboratories and centers for disease control. They fight to combat both local outbreaks of food poisoning and also such major international threats as AIDS and avian flu.

100. How do surgeons learn?

This photograph shows the anatomy theater at the University of Padua in northern Italy, originally built in 1594. At the time Padua was the world's top medical school, and students went there from far afield, including Englishman William Harvey (1578–1657), who arrived in 1600. He must have watched dissections in this very theater, and later worked out how blood circulates around the body, writing the definitive book about it in 1628.

For me, however, the real medical hero of Padua was a Belgian anatomist named Andreas Vesalius (1514–64), who was maddened by the current teaching methods. The learned professor, having no intention of getting his hands dirty, would stand at one end of the room and read out instructions in Latin from the translated text of the ancient Greek physician Galen (c. 129–c. 199 A.D.), while the lowly surgeon, who usually did not understand Latin, hacked incompetently at the corpse. His job was made more difficult by the fact that Galen had rarely dissected any human cadavers; most of his descriptions referred to pigs, monkeys and dogs.

Persuading a judge to let him have the bodies of executed criminals, Vesalius began to do his own dissections, which was revolutionary—people came from miles around to see the learned professor rolling up his sleeves and brandishing the knife. He was one of the first-ever champions of hands-on science. More important, he drew what he saw, and produced the most beautiful book on the anatomy of the human body—*De humani corporis fabrica*. All the illustrations were stunning engravings, carved in pear wood by the finest artists from Venice.

Once his book was published in 1543, the anatomists knew what they were doing, and in this theater students like Harvey leaned over the carved walnut rails on all six tiers, so crowded and intimate that even in the top row the students were only 13 feet (4 m) from the corpse as Vesalius's successors explained what they were doing.

According to legend, cutting up dead bodies was frowned upon, so they always kept a partly cut-up pig on a second table in the basement. If the authorities appeared at the door, they would be briefly delayed while the human cadaver was removed and replaced by the pig.

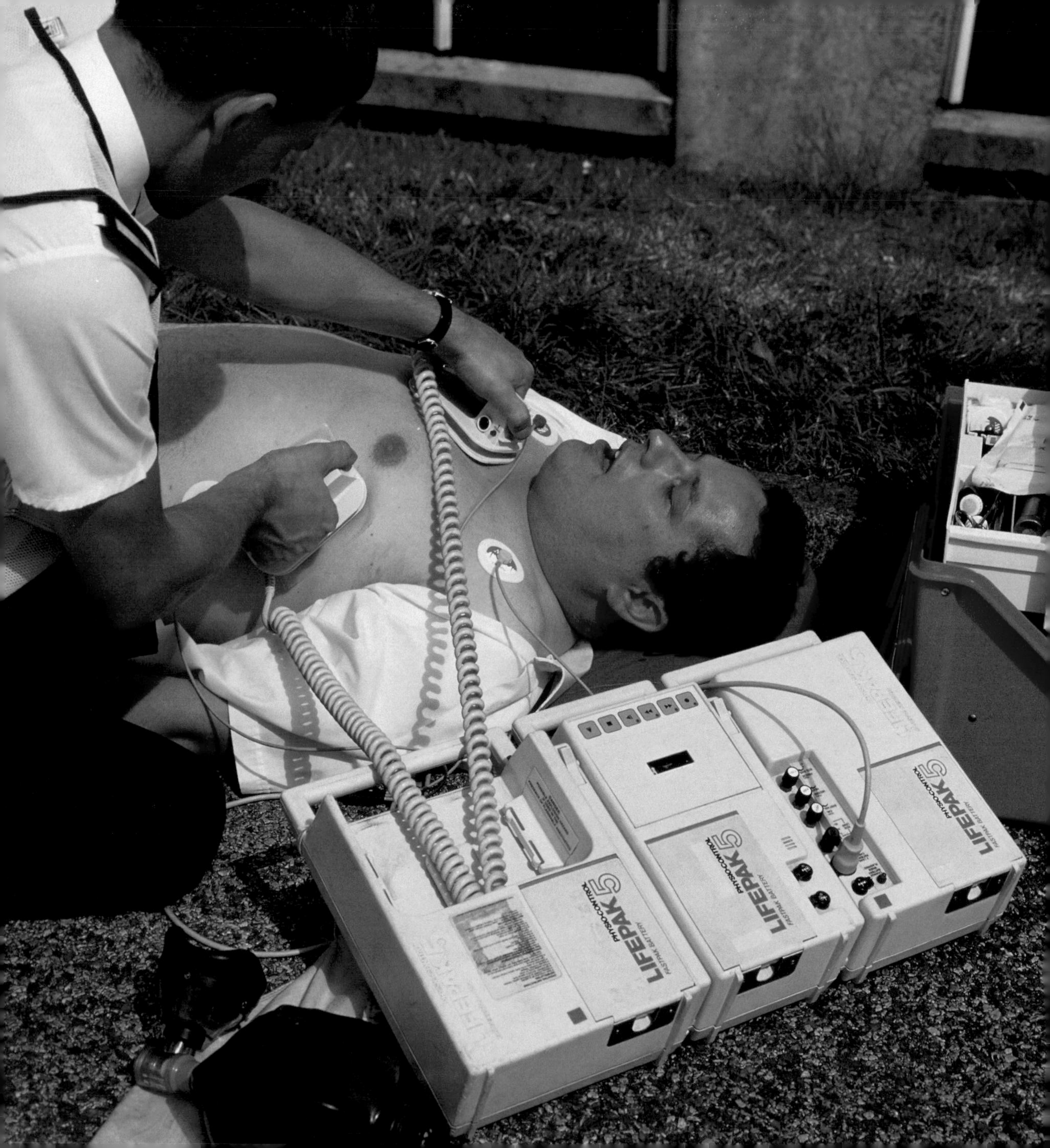
PHYSIO-CONTROL
LIFEPAK 5
FASTPAK BATTERY
PHYSIO-CONTROL
LIFEPAK 5
FASTPAK BATTERY
PHYSIO-CONTROL
LIFEPAK 5
FASTPAK BATTERY

101. What happens in a cardiac arrest?

The heart is a bag of muscle, about the size of your fist, that pumps blood around your body. It has a boring and repetitive job, beating about once a second for your whole life. When it stops you are dead. Well, not quite; if the heart can be started again you may survive.

What actually triggers the heart to beat is an electrical signal that comes from your autonomous nervous system; you don't have to think about it.

DID YOU KNOW?

When you get scared, excited or are furiously exercising, your heart beats faster—up to 200 beats per minute for an athlete in competition.

Occasionally, the signaling goes wrong, which may cause an irregular heartbeat or, in an extreme case, a cardiac arrest: the heart stops completely or goes into a useless fluttering called ventricular fibrillation.

A person with irregular heartbeat may be fitted with a pacemaker. This is a small electronic package that is placed under the skin and delivers a tiny electric shock to the heart every second or so; in other words, it takes over the job of the autonomous nervous system in sending these signals. Because it sends the signals at regular intervals the heartbeat is once again regular.

DID YOU KNOW?

Cardiac arrest, when the heart stops beating properly, is not the same as a heart attack. In a heart attack, the blood flow to the heart muscle is interrupted, often by a chunk of plaque caused by build-up of cholesterol, and the heart muscle itself is damaged.

Cardiac arrest may be caused by shock or overload of the system. It is most common with fat elderly men. The victim will die within a few minutes unless the heartbeat is restarted. The best hope is to shock the heart muscle back into normal functioning. In this photograph, a paramedic is attempting to restart the patient's heart by using a portable defibrillator to apply a massive electric shock to the chest.

In the absence of a defibrillator, the best hope is to massage the heart; start with a thump at the top of the rib cage, then lean on the ribs every second for 15 seconds, give two breaths of mouth-to-mouth resuscitation, and lean on the ribs again.

technical details

film

All the photographs in this book were taken on positive transparency film, mostly Fujichrome Provia 100 ISO, or occasionally 400. Some of the earlier ones were taken on Kodachrome 25 or 64.

I enjoy using a digital camera for fun, but am advised that the quality does not yet match that of film.

cameras

Most of these photographs were taken with medium-format cameras. My workhorse is a Mamiya RB67, a lovely instrument that I bought secondhand 20 years ago, and it has not yet let me down. It is simply a box with a lens on the front—usually a 140 mm macro lens, sometimes with extension tubes—a film holder at the back, and a viewfinder on top. All manual and no battery to fail. It takes 120 roll film, giving 10 transparencies $2^1/_4$ x $2^3/_4$ inches (6 cm x 7 cm).

My previous workhorse, still much used today when I need speed and mobility, is an Olympus OM2n 35 mm camera, with Olympus Zuiko lenses of 16 mm, 28 mm, 50 mm, 100 mm, and especially a 50 mm macro lens with extension tubes.

Other cameras used for pictures in the book include a Horseman technical 5 x 4 monorail camera with a 180 mm Rodenstock lens, a Pentax 67, a Mamiya 7 and a Rolleiflex.

For the studio work I have built a Dexion (metal shelving) frame 2 feet square (6 m sq.) and from floor to ceiling. On this frame I bolt or clamp the flashguns, reflectors, such apparatus as Mazof trigger, burette, or dart-dropping machine, the Plexiglas background and the subject. The camera is mounted either on the frame, on an extension of it or on a separate heavyweight camera stand. Thus I avoid as far as possible having tripods to trip over.

flash equipment

I take many of my close-up photographs in a darkened room, by opening the camera shutter, firing the flash, and then closing the shutter again.

I have a number of flash guns, which have a flash duration of around one millisecond ($^1/_{1,000}$ of a second).

When I want to freeze rapid motion, I need high-speed flash—a flash that is over in a much shorter time—and I use mainly Metz hammerhead guns in W (Winder) mode, which gives a flash duration of about 100 microseconds (or $^1/_{10,000}$ of a second). I often fire this either with a Mazof trigger or with triggers designed and supplied by Loren Winters (www.hiviz.com). High-speed flash is much weaker than normal flash, and therefore I took many of the high-speed pictures on fast (400 ISO) film.

For the strobe photographs on pages 40, 131 and 134 I used a Martin Atomic 3000 lamp.

backgrounds

For most of my close-up photographs I use an artificial background. For black I use black velvet, 6 feet (2 m) back if possible. For white or colored backgrounds I use a sheet of 3 mm opal Plexiglas perhaps 20 inches (50 cm) behind the subject and lit from the far side.

Lighting the background with a single flashgun pointing at the camera sometimes gives a bright area in the center of the picture with darker edges (such as page 152); so often I use a pair of flashguns, right and left, pointing inward at about 45 degrees. This gives more even light. When I want color I stick color gels over the flashguns, which allows two-tone backgrounds.

individual photographs

Those not listed below were taken either by natural light or with a simple flash setup.

Page 11 I used Meccano to make a dart-dropping machine, which I fixed to the top of my frame. The balloon was taped to a piece of wire protruding through the background. The balloon was lit with high-speed flash. I opened the camera shutter, pulled the string to drop the dart. An infrared beam shone through the top of the balloon and activated a Mazof trigger to fire the flash just after the balloon had begun to burst.

Page 19 The feather was lying on a horizontal sheet of glass and shot from below, through the glass. This is better than shooting from above, which can produce annoying reflections of the subject in the glass.

Page 20 I persuaded some glass-making friends to break the bottom off the champagne bottle. I clamped the top end to my frame, squirted soda water up through the neck from a squeezy bottle with one hand, and pressed the cable release with the other; the flash was synched with the shutter. The cork was stuck on the end of a piece of wire (a straightened coat hanger) poking through the background. The picture would have been more convinc-

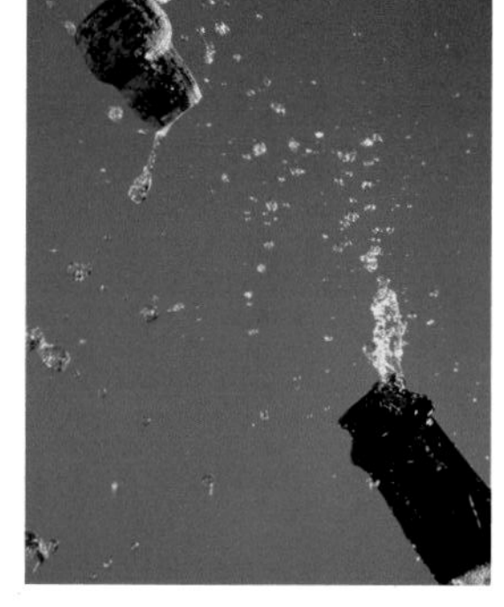

ing if I had devised a way of getting some movement into the cork, although in real life the cork travels many meters before any liquid comes out of the bottle.

Page 40 This is a double exposure. First I photographed the branch, then I removed it and for the second exposure dropped the apple while lighting it with a strobe lamp.

Page 47 The water was dripping from a burette into a large black plastic gardener's seed tray containing about an inch (3 cm) of water. The Plexiglas background was about 6 inches (15 cm) behind the subject, above the water surface. It is lit from behind with a single flashgun, and the lurid colors are produced by several strips of colored gel, taped to the back of the Plexiglas.

The falling drop goes through an infrared beam, and so activates the Mazof trigger, which fires the flash after a delay that can be varied. By adjusting the delay I can choose exactly what stage of the drop to photograph.

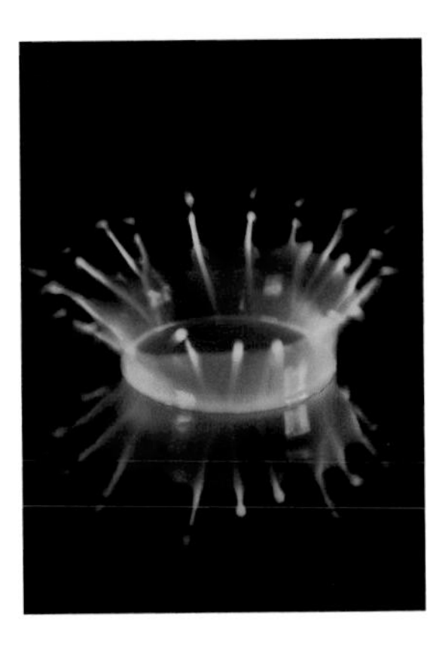

Page 48 The milk drop has fallen onto a little pool of milk on a sheet of black Plexiglas. I used the same burette and trigger system as for the water drop on page 47, but for this picture I lit the liquid from the sides, using blue and green gels, rather than from behind.

Also, I used a normal (rather than high-speed) flash, with the result that each of the droplets in the crown appears to have a point. The flash starts very bright and slowly dims, which means that each drop becomes a streak, and as the flash dims the image seems to get smaller. Each point shows how far the droplet travels in about a thousandth of a second. This is why I need a high-speed flash to freeze the movement, as on page 53.

Page 51 For this picture I made a little rain machine by cutting and gluing old plastic tubes to give me half a dozen streams of water from a trough. I used a high-speed flash.

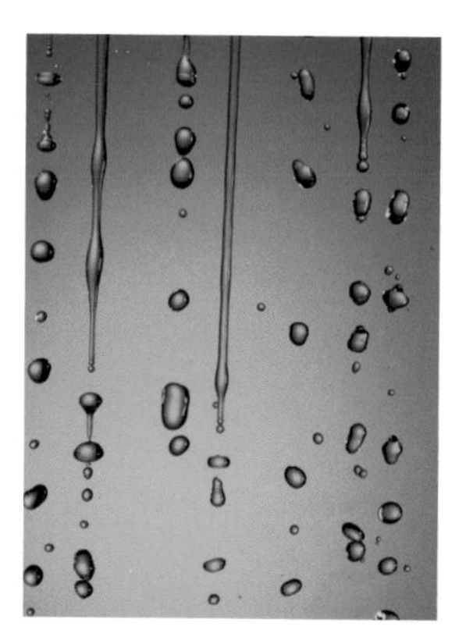

Page 61 I made the whirlpool in a Plexiglas cylinder within a fish tank full of water, using a magnetic stirrer—the stirrer magnet is just out of shot at the bottom of the picture.

Page 53 High-speed flash—note the difference from the milk crown on page 48.

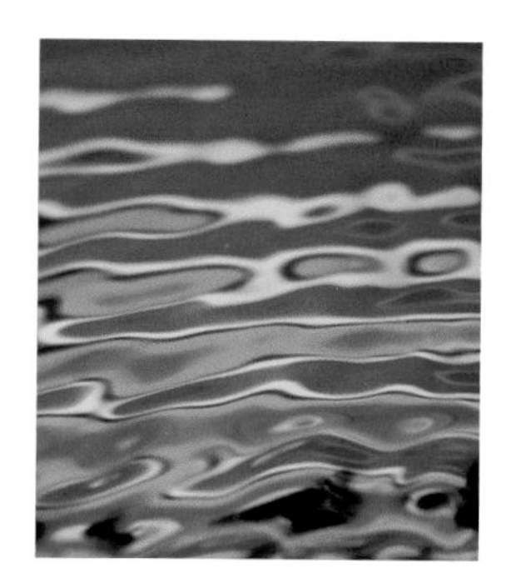

Page 65 This is only water, in a large black tray, which I vibrated with a cordless electric drill. The colors are from multiple strips of gel.

Page 57 High-speed flash.

Page 72 This is the grandson of a former neighbor.

Page 75 I borrowed a tesla coil from a school and grounded all three wires in the lead from the plug. In darkness I first photographed the plug using flash, then left the shutter open for a few minutes while I held the end of the tesla coil just close enough to each pin in turn, and for long enough to accumulate many sparks. The tesla coil does not show up because it was not lit and I kept it moving.

Page 76 I borrowed a wig from the makeup department at the TV company where I was working, and found it just fitted on the top dome of a van der Graaf machine that I borrowed from a physics department.

Page 81 A few seconds later the limelight became much too bright to look at or photograph.

Page 84 This is sunlight, coming through a narrow slit into a darkened room.

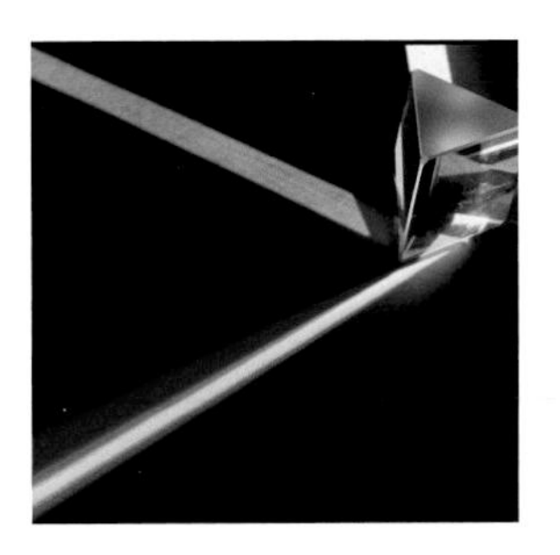

Page 91 The fiber optics are lit with flash and gels of different colors from the side and through from the other ends.

Page 92 The difficult thing about this photograph was getting accurate focus on the eye in the dark; I could not illuminate her eye or the pupil would have contracted.

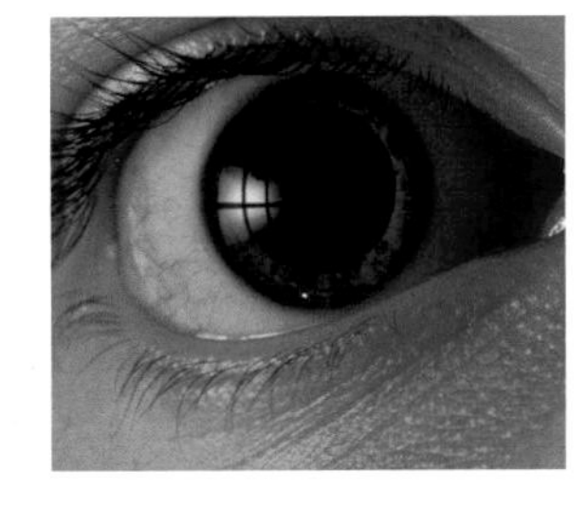

Page 95 To shoot the moon I followed a helpful book by Michael Covington, *Astrophotography for the Amateur*, (Cambridge University Press, 1999). My sequence of the eclipse was eventually foiled by cloud.

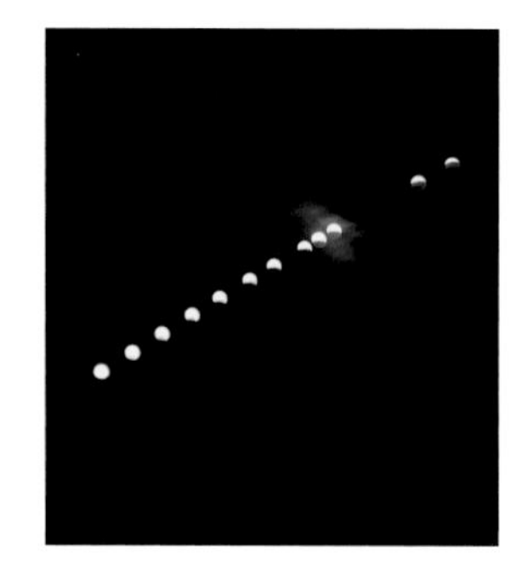

Page 111 High-speed flash.

Page 128 I used my Rolleiflex for the harmonograms because it takes square pictures which fit the patterns well, because it is lightweight, and because it is easy to mount pointing downward.

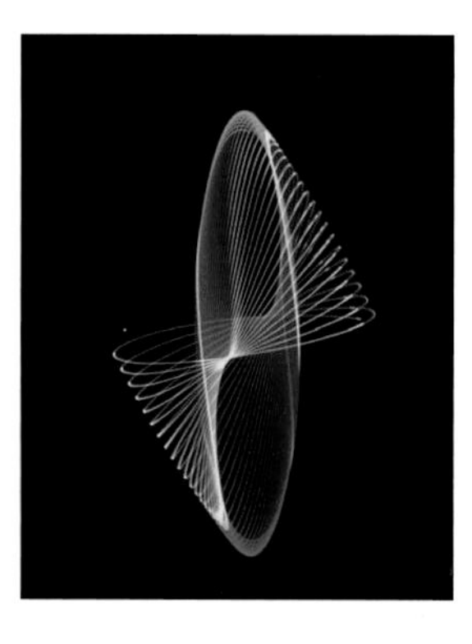

I built my harmonograph from two pendulums, each made of 5 feet (1.6 m) of 40 millimeter plastic plumbing pipe, suspended from a universal joint. One pendulum was hung from the ceiling, with a camera mounted at the bottom, pointing downward. The other was hung 35 inches (90 cm) lower from a frame 24 inches (60 cm) to one side, with a stiff wire projecting horizontally at the bottom from its heavy bob. On the end of this wire was fixed an LED, pointing upward, so that at rest it pointed straight up at the lens. The room was blacked out, both pendulums were set swinging, the camera's shutter was opened (on B), and after a few seconds, the LED was switched on. The pendulums were allowed to swing for about a minute, then the LED was switched off and the shutter closed. See also "harmonograph" on www.google.com.

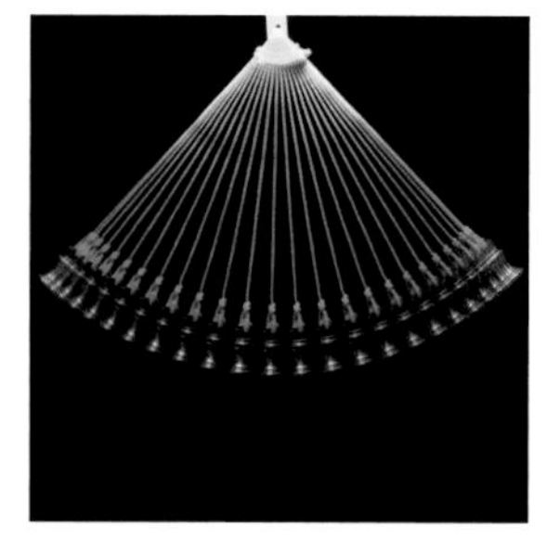

Page 131 The pendulum was lit with the strobe lamp for a half-second exposure.

Page 134 The ball was lit with the strobe light, running on "blinder" mode at 50 Hz (flashes per second).

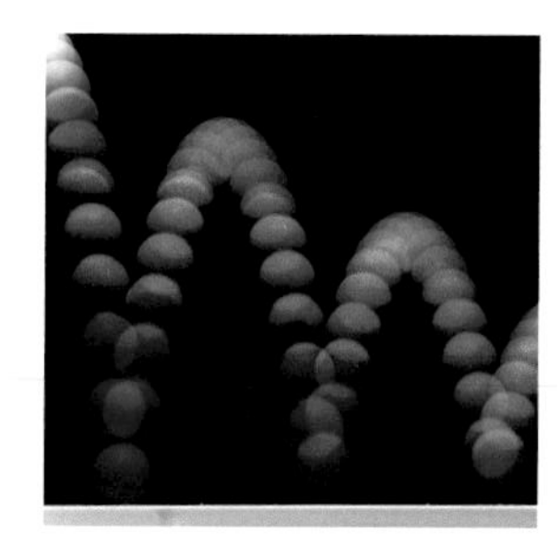

Page 189 Spiders' webs are surprisingly difficult to photograph, probably because the strands are so fine. The best way I have found is to get flash guns close on each side, but that tends to overlight the spider (as here) and anything else in the picture, such as twigs and leaves.

Page 159 This is a double exposure; I moved the flower, refocused and changed the lighting in between.

Page 202 These abstract pills are a double exposure.

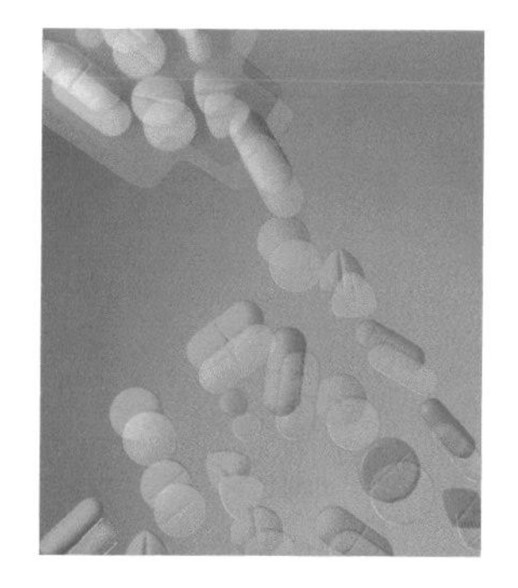

Page 186 I wanted to show tadpoles at various stages of development, so I took a few from the pond and kept them for a day or two in pond water in the fridge, to slow them down. I then put all the tadpoles in pond water in a shallow clear plastic tray and lit them diagonally from below.

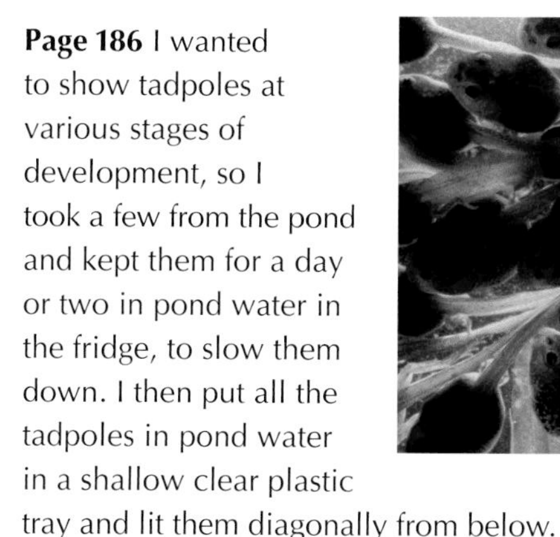

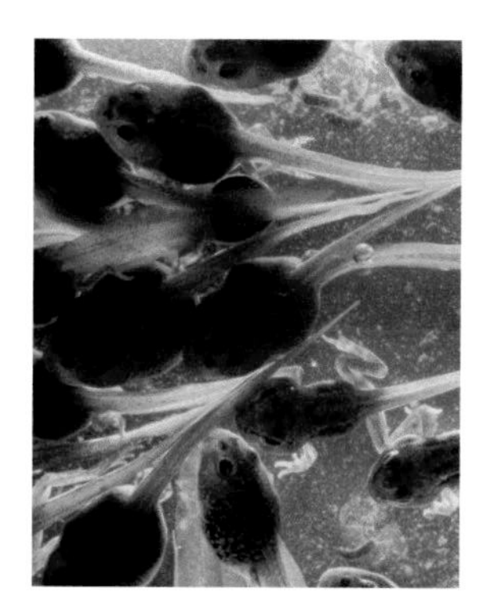

Page 210 This is a simulation, by helpful volunteers from the West Yorkshire Ambulance Service. The "victim" had not had a cardiac arrest and did not receive an electric shock.

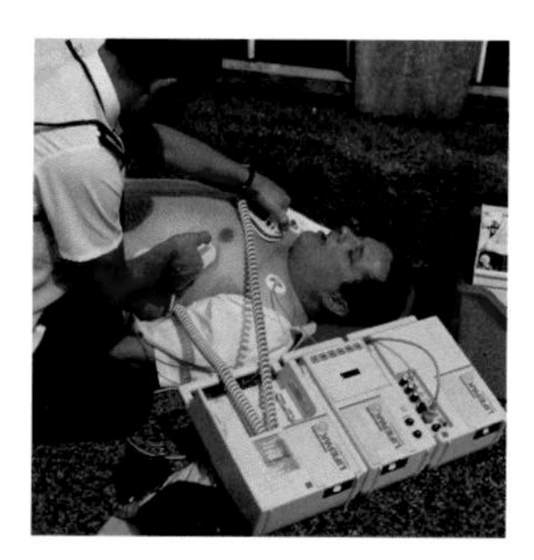

acknowledgments

The seed for this book was sown by my agent, Mandy Little, who liked my photographs, and began telling me that I should write a book about them. She told other people too, and at Ebury Publishing, Jake Lingwood took up the gauntlet; then Carey Smith and Sarah Lavelle shaped and moulded the book, and coped with my tantrums.

The photographs in the book are all mine, but the Science Photo Library hold many of them in their files and supplied scans for the book, for which I thank them. I wish particularly to thank Rosemary Taylor, who has encouraged me for 22 years and has taught me more about photography than anyone else. Often I would send her a hundred pictures for the library, and back would come 97 with little yellow stickers carrying such helpful comments as, "Why don't you try getting it in focus?"

I am most grateful to the following scientists for reading parts of the text and picking up the worst of my mistakes: Professor Sir Michael Berry, Don Cameron, Dr. Hermione Cockburn, Dr. Barbara Dunning, Dr. David Jones, Juliet Jopson, Dr. Marty Jopson, John Martin, Chris O'Toole, Dr. Janet Sumner, Professor Tom Troscianko and Professor Lewis Wolpert.

I thank my family for putting up with many tedious photographic exploits, and especially with my extra grumpiness during the last couple of months.

index